T0137239

Digital Storage in Consumer Electronics

Thomas M. Coughlin

Digital Storage in Consumer Electronics

The Essential Guide

Thomas M. Coughlin
Atascadero, CA, USA

ISBN 978-3-319-88860-6 ISBN 978-3-319-69907-3 (eBook)
https://doi.org/10.1007/978-3-319-69907-3

Softcover reprint of the hardcover 2nd edition 2017

Printed on acid-free paper

This Springer imprint is published by Springer Nature
The registered company is Springer International Publishing AG
The registered company address is: Gewerbestrasse 11, 6330 Cham, Switzerland

This book is dedicated to my mother, Connie, who taught me to keep working until a project is done. It is also dedicated to my family, Fran, Will, and Ben, for putting up with my time away from them to work on this book.

Preface to the Second Edition

A lot of things have changed in consumer electronics and the technology that enables them in the 10 years since I wrote the first edition of this book in 2007. In 2007, the first iPhone was released, and small hard disk drives were being used to store music on portable music players. Today, all mobile devices use flash memory for storage, and consumers are wearing digital watches and even digital rings. Consumers are using the Internet to stream videos and store their content or for accessing application from the cloud. As a result of these developments, the second edition of this book had undergone significant modifications.

In all the chapters, content data rates and capacities have been updated including today's 4K as well as tomorrow's 8K content. Likewise, we have updated many product schematics and teardown examples thanks to companies such as Texas Memory and iFixit.

Chapter 1 was modified to discuss current consumer electronic trends, and an expanded analysis of digital storage hierarchies gives a better idea why designers may use one form of digital storage or another for a consumer device or application. Chapters 2, 3, and 4 were updated to cover current developments with hard disk drives, optical disks, and flash memory. Chapter 5 was updated to include network video recording and the rise of Internet-based Video on Demand as well as network DVRs. We also look at the development of smart TVs and new generations of set-top boxes as well as Internet network boxes such as Apple TV and discuss media centers.

Chapter 6 was expanded to include digital storage and memory for modern automotive applications such as infotainment and advanced driving and autonomous driving as well as mobile consumer products such as music players, smart phones, tablet computers, cameras, GPS map devices, smart watches, and even smart rings. Chapter 7 is a new chapter looking at enabling technologies for mobile consumer devices including high-resolution display technologies, advances in mobile power technology, developments in consumer metadata, and applications with machine learning and other forms of artificial intelligence.

Chapter 8 updates the bill of materials costs for consumer electronic products including digital storage. It also broadens the discussion of object storage and its use in consumer applications. Chapter 9 expands the home networking discussion from the first edition, in addition to the growth of cloud services including storage as well as the consumer applications using the Internet of Things. Chapter 10 has an updated estimate of consumer storage requirements and several updates in projections for digital storage devices used in consumer electronics. Chapter 11 updates the reference and appendix information, including current companies providing storage products for consumer electronics.

Atascadero, CA, USA Thomas M. Coughlin

Acknowledgments

An undertaking such as this book requires lots of support from one's peers as well as one's family. Many folks reviewed and suggested content for the various chapters. In Chap. 2, Dick Zech and Brian Berg provided significant input on the optical drive section, while Jim Handy provided excellent illustrative figures and other materials to describe the action of flash memory devices. Brian Berg also reviewed the hard disk drive section. Rick Wietfeldt reviewed Chaps. 5 and 6 in the first edition. Bert Haskell also reviewed Chaps. 5 and 6 in the first edition and gave me guidance in determining costs for consumer products. Michael Willett reviewed the Trusted Computing Group material in Chap. 7. Tips and hints came from many sources.

Overall reviewers of the book include Tom Clark and Brian Berg on the first edition. Thank you for all your suggestions and input.

Thanks to Texas Instruments for schematics and many sources of information on various consumer devices. Also thanks to Peek Inside, iFixit, and other sources for teardown images and information.

All good ideas and suggestions included in these pages are inspired by these folks and many others. I take responsibility for all the mistakes and omissions.

Contents

About the Author

Thomas M. Coughlin is the founder and president of the consulting firm Coughlin Associates. Tom has over 35 years of experience in the data storage industry as a working engineer and high-level technical manager. In addition to regular technical and management consulting projects, he is the publisher of many reports covering technology and applications for digital storage devices and systems and writes a regular blog on digital storage topics for Forbes.com. He has many published papers, reports, and articles and is a frequent contributor to the *IEEE Consumer Electronics Magazine* where he is also an associate editor. He has six patents on magnetic recording and related technologies. Tom is the founder and organizer of the annual Storage Visions Conference and the Creative Storage Conference. Tom is an IEEE Fellow and has held many volunteer positions in the IEEE, including President-elect of IEEE-USA, director of IEEE Region 6 and vice president of Operations and Planning for the IEEE Consumer Electronics Society. He is also a member of APS, AVS, SNIA, TCG, and SMPTE. Tom has a BS in physics and an MSEE from the University of Minnesota and a PhD in electrical engineering from Shinshu University in Nagano, Japan. For more information on Coughlin Associates, go to www.tomcoughlin.com. For more information on the Storage Visions Conference, go to www.storagevisions.com. For more on the Creative Storage Conference, go to www.creativestorage.org.

Chapter 1
Introduction

1.1 Objectives in this Chapter

- Look at consumer electronic trends that drive demand for digital storage.
- Develop an appreciation for the role of digital storage in the growth of consumer electronics.
- Understand the key role of product price in the growth of consumer products for most markets and how this is impacted by the cost of the digital storage used.
- Get an initial exposure to the development of standard consumer device functions and how these could be integrated more tightly with the digital storage devices.
- Present some rules for the design of digital storage into consumer devices in order to create more successful products.
- Review the role of memory in processor execution as well as mass storage.
- Develop the concepts of a digital storage hierarchy for mobile and static consumer applications.
- Demonstrate the advantages of using multiple storage devices in consumer products creating hybrid storage products.

1.2 The Growth of Consumer Electronics

Consumers use a number of electronic devices, some mobile and some relatively fixed. In recent years, many popular devices have migrated from larger fixed products, like desktop computers, to mobile computers, tablet computers, and smart phones. Some consumer markets, like the USA, may have saturated their need for new devices, with sales instead being replacements for older devices. Many other consumer markets are still growing with consumers getting their first consumer version of a consumer electronic device. In these markets, these devices are more likely mobile than fixed.

T.M. Coughlin, *Digital Storage in Consumer Electronics*,
https://doi.org/10.1007/978-3-319-69907-3_1

Table 1.1 US consumer personal device unit shipments (M) from 2016 through 2021[1]

Table 2: United States Consumer Personal Device Unit Shipments (M)								
PCs	2016	2017	2018	2019	2020	2021	CAGR	Trend
Desktop PC	5.1	4.9	4.1	3.4	2.8	2.3	-14.3%	
AGR	*-12.4%*	*-4.2%*	*-15.5%*	*-17.1%*	*-18.0%*	*-16.3%*		
Mobile PCs	20.1	18.5	19.7	20.8	21.8	23.0	2.8%	
AGR	*-8.1%*	*-8.1%*	*6.7%*	*5.7%*	*4.7%*	*5.7%*		
Total PCs	25.1	23.3	23.8	24.2	24.6	25.4	0.2%	
AGR	*-9.0%*	*-7.3%*	*2.1%*	*1.8%*	*1.5%*	*3.2%*		
Tablets								
Detachable	5.4	4.7	5.3	5.8	6.2	6.9	5.0%	
AGR	*-15.1%*	*-12.6%*	*11.2%*	*9.1%*	*8.0%*	*11.2%*		
Slate	35.3	29.4	23.7	19.2	15.4	12.4	-18.9%	
AGR	*-12.0%*	*-16.7%*	*-19.2%*	*-19.1%*	*-19.9%*	*-19.4%*		
Total Tablets	40.7	34.1	29.0	25.0	21.6	19.3	-13.8%	
AGR	*-12.4%*	*-16.2%*	*-14.9%*	*-14.0%*	*-13.5%*	*-10.6%*		
Total Computers								
Total Computers	65.8	57.4	52.8	49.2	46.2	44.7	-7.5%	
AGR	*-11.1%*	*-12.8%*	*-8.0%*	*-6.9%*	*-6.1%*	*-3.2%*		
Mobile Phones								
Stardard Phone	20.1	18.0	14.8	10.8	7.9	6.2	-21.1%	
AGR	*-27.5%*	*-10.6%*	*-17.6%*	*-27.2%*	*-26.3%*	*-22.4%*		
SmartPhone	170.7	178.8	188.5	194.7	199.2	207.9	4.0%	
AGR	*4.7%*	*4.7%*	*5.4%*	*3.3%*	*2.3%*	*4.4%*		
Total Mobile Phones	190.8	196.8	203.3	205.5	207.1	214.0	2.3%	
AGR	*0.0%*	*3.1%*	*3.3%*	*1.1%*	*0.8%*	*3.3%*		
Total Devices								
Total Devices	256.7	254.2	256.1	254.7	253.3	258.7	0.2%	
AGR	*-3.1%*	*-1.0%*	*0.7%*	*-0.5%*	*-0.5%*	*2.1%*		
∞∫Δ Daniel Research Group © (2017)								

Table 1.2 US consumer personal device unit shipments from 2016 through 2021[1]

Table 13 United States Consumer Personal Devices								
	2016	2017	2018	2019	2020	2021	CAGR	Trend
Unit Shipments (K)	256,672	254,216	256,114	254,715	253,314	258,729	*0.2%*	
AGR	*-3.1%*	*-1.0%*	*0.7%*	*-0.5%*	*-0.5%*	*2.1%*		
Revenue ($M)	74,657	73,389	73,270	72,079	70,094	70,126	*-1.2%*	
AGR	*-11.8%*	*-1.7%*	*-0.2%*	*-1.6%*	*-2.8%*	*0.0%*		
Average Price ($)	290.87	288.69	286.08	282.98	276.71	271.04	*-1.4%*	
AGR	*-9.0%*	*-0.7%*	*-0.9%*	*-1.1%*	*-2.2%*	*-2.0%*		
Installed Base (K)	584,172	589,446	575,601	573,994	580,159	570,764	*-0.5%*	
AGR	*6.4%*	*0.9%*	*-2.3%*	*-0.3%*	*1.1%*	*-1.6%*		
Installed Base Age (Y)	2.47	2.54	2.58	2.58	2.57	2.55	*0.6%*	
AGR	*2.7%*	*2.8%*	*1.6%*	*0.0%*	*-0.5%*	*-0.5%*		
Replacement Cycle Length (Y)	3.64	3.37	3.13	3.24	3.35	3.13	*-3.0%*	
AGR	*8.6%*	*-7.4%*	*-7.0%*	*3.4%*	*3.3%*	*-6.5%*		
Units per Households (#)	4.65	4.64	4.48	4.42	4.42	4.31	*-1.5%*	
AGR	*5.3%*	*-0.1%*	*-3.4%*	*-1.4%*	*0.0%*	*-2.6%*		
	2016	2017	2018	2019	2020	2021	Change	Trend
Market Penetration (%)	*99.9%*	*100.0%*	*100.0%*	*100.0%*	*100.0%*	*100.0%*	*0.0%*	
∞∫Δ Daniel Research Group © (2017)								

Table 1.1 shows projected shipments of various consumer electronic devices in the USA from 2016 through 2021. These projections show a continued decline in desktop PCs, some increase in mobile PCs, a general decline in tablet computers (except detachable units), an increase in smart phones, and a decline in traditional cell phones. Table 1.2 shows summary information on all PCs, tablets, and cellular phones from 2016 through 2021. This includes total unit shipments, revenue, installed base, installed base age, average replacement time, and total number of units per household.[1] Figure 1.1

[1] DRG (Daniel Research Group) United States Computing and Telecommunications Personal Device Market Forecast: 2017–2021, July 2017 Update, http://www.danielresearchgroup.com/

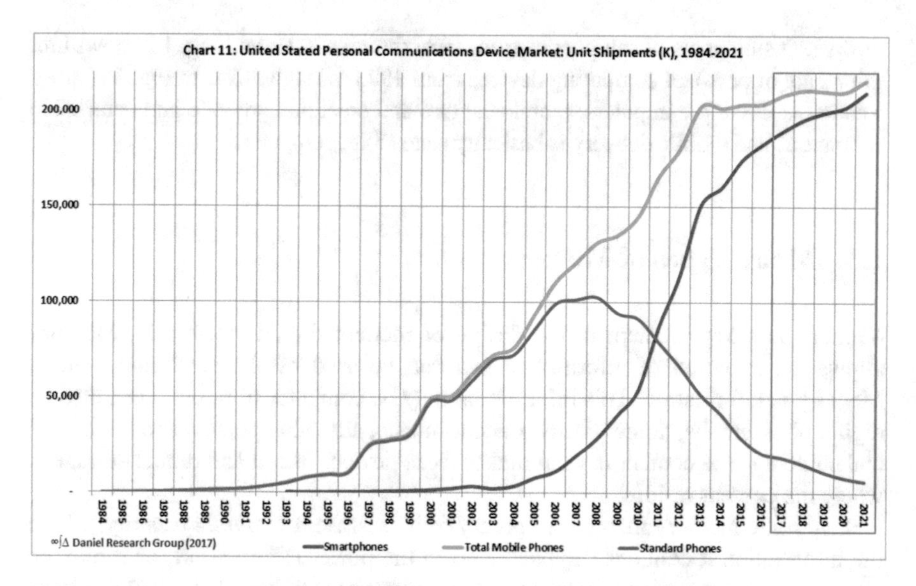

Fig. 1.1 US personal communications device market unit shipments (K) from 1984 to 2021[1]

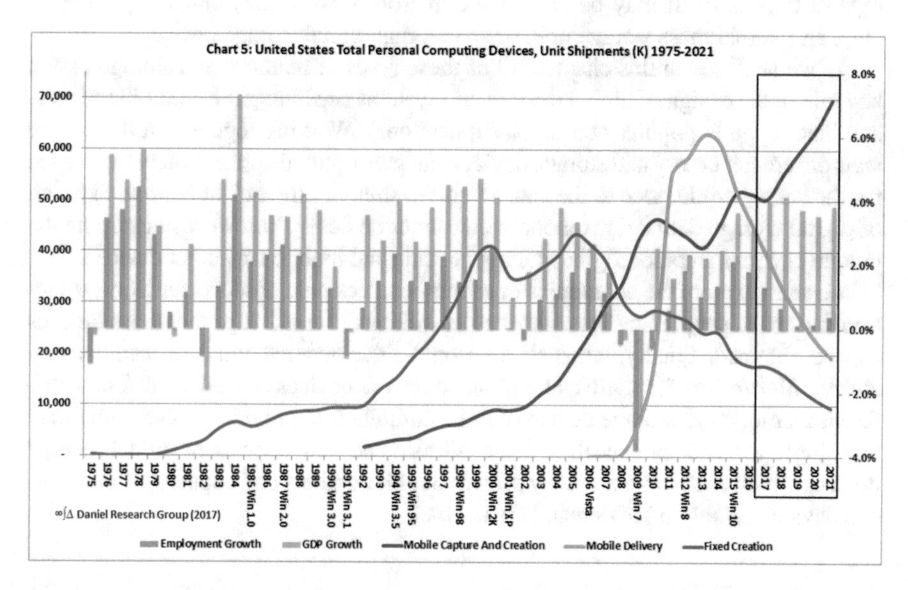

Fig. 1.2 US personal computing device market unit shipments (K) from 1975 to 2021[1]. The growing number of consumer electronic devices in the home and around our bodies is driving our needs for digital storage and memory to capture content, read news, watch videos, and carry on all the applications and social networking that are popular today

shows unit shipments of cell phones from 1984 through 2021, and Fig. 1.2 shows unit shipments of personal computing devices from 1975 through 2021. In Fig. 1.2, fixed creation is desktop computers, mobile capture and creation is mobile and detachable computers, and mobile delivery is basically slate tablet computers.

1.3 Many Types of Memory

Memory is a key element in the design of modern digital electronics. Memory allows the retention of information that can be used by the electronic system. Memory is information. This information may be computer files, but it can also be digital photographs, home videos, movies, music, and other personal and commercial contents. The content is what makes these devices useful and digital storage is where the content resides.

Electronic memory may be temporary as is the case for *volatile memory*, where the information it contains disappears when the power is turned off, or it may be long-term memory, or *nonvolatile memory*. Memory may be fast or slow and have low-power consumption or high power consumption, and it may be inexpensive per byte or expensive. It may be part of the microprocessor electronics, a peripheral chip, an internal mass storage unit, or an external digital storage device.

As we shall see in this chapter, all of these types of memory and storage play a key role in the design of digital devices and systems (including the cloud-based storage that supports popular consumer applications). Working together in the proper way, different memory and storage devices can give optimal application performance for the best possible price to the consumer. We shall briefly look at expected growth of digital storage demand for modern consumer devices. Then we will examine the unique environments for digital storage in static and mobile consumer devices.

We shall explore the economics of consumer electronic products and look at factors that make people buy these devices as well as different ways that these devices can be obtained. Finally, we shall develop a finer understanding of *memory and storage hierarchies* for static and mobile consumer devices as well as look how different memory and storage devices can be combined to provide greater value than any single technology can offer. We shall also see how dependent digital storage devices are on each other and how growth in unit numbers and capacity of one storage device can help drive demand for another.

1.4 Growth in Digital Content Drives Storage Growth

Demand for digital storage is driven by the growth in personal family content, by the growing number of entertainment devices in and around the home that require digital storage, by the increasing resolution of personal and commercial entertainment content, and by the growth in the number of channels that people can use to access all types of content.

Most consumers have digital still and video cameras, either as stand-alone devices or in convergence devices such as cell phones. Content creation devices such as digital still and video cameras are becoming a common *standard application* that can be built into consumer devices, like smart phones. In addition to being ubiquitous, increasing camera resolutions make digital storage demands higher with time.

Families today have many ways to enjoy content. No longer must they be tied to a schedule for programs on the television. Content can be recorded from broadcast, cable, or satellite broadcasts on a digital video recorder or recovered from a network DVR to play back later. Audio or video content can be downloaded from the Internet and listened to on a variety of static and mobile devices such as MP3 players and smart phones. This content may be stored on personal computers and possibly backed up to external storage devices.

In a number of homes, content may be stored in a network storage device and made available through a wired or wireless network to various static and mobile devices throughout the home. Automobiles are increasingly using digital content for entertainment and navigation purposes. In the future, all of these mobile and static devices may be part of a far more comprehensive home storage network architecture.

Content sharing and access is growing with time (note the popularity of YouTube). Content can be streamed or downloaded from the Internet and watched now or later, it can be brought down to a cell phone through the mobile phone network, and it can be shared with everyone via the many growing social networking web sites.

In all cases, the resolution of the content that people want is increasing. But to receive this content, it must be compressed. Content is compressed for two basic reasons. First, it is compressed to get through slow network connections and to speed up downloading or streaming. Second, it is compressed to conserve digital storage space required. *Compression* can be lossy or lossless. Lossy compressed content cannot be reconstructed to its original resolution, whereas lossless content can be fully reconstructed using decompression technology.

With increasing Internet connection speeds and the low cost of digital storage capacity, we are likely to see changes in the resolution of content people may want. For instance, MP3 files are a lossy compressed format where up to 90% of the original content file is lost during compression. The compression is done with a very fine human hearing compression model in a noisy environment and with less than optimal acoustic equipment you probably can't tell that there is content missing. With faster Internet speeds and low-cost storage, there is beginning to be a pronounced shift from lower-quality MP3 formats to less compressed or even lossless compressed music formats.

Table 1.3 shows some projections for a number of digital photographs, music files, and video files for different resolutions and device capacity that a user might require sometime in the not so distant future. The gray areas for each media format show an estimate of the desired unit content ranges. These are based on an estimate of how much content an individual would like to have available on-demand from a local (e.g., portable device). Note that the requirements are much different if much of this content is stored in the cloud:

Table 1.3 Media units vs. storage capacity for various resolution photos, music, and video files

Object Size (MB)	1	2	4	150	667	4,160
Capacity (GB)	**4 MP Photos**	**8 MP Photos**	**MP3 Songs**	**HiD Song**	**VGA Video**	**DVD Video**
5	5,000	2,500	1,250	33	7	1
10	10,000	5,000	2,500	67	15	2
15	15,000	7,500	3,750	100	22	4
20	20,000	10,000	5,000	133	30	5
25	25,000	12,500	6,250	167	37	6
30	30,000	15,000	7,500	200	45	7
35	35,000	17,500	8,750	233	52	8
40	40,000	20,000	10,000	267	60	10
45	45,000	22,500	11,250	300	67	11
50	50,000	25,000	12,500	333	75	12
55	55,000	27,500	13,750	367	82	13
60	60,000	30,000	15,000	400	90	14
65	65,000	32,500	16,250	433	97	16
70	70,000	35,000	17,500	467	105	17
75	75,000	37,500	18,750	500	112	18
80	80,000	40,000	20,000	533	120	19
85	85,000	42,500	21,250	567	127	20
90	90,000	45,000	22,500	600	135	22
95	95,000	47,500	23,750	633	142	23
100	100,000	50,000	25,000	667	150	24
105	105,000	52,500	26,250	700	157	25
110	110,000	55,000	27,500	733	165	26
115	115,000	57,500	28,750	767	172	28
120	120,000	60,000	30,000	800	180	29
125	125,000	62,500	31,250	833	187	30

- Up to 20,000 photo images
- Up to 10,000 songs
- Up to 100 movies

This chart can give us an estimate of the acceptable size of storage for various applications at various resolutions; the following are some examples:

- A 4-megapixel photo viewer with 20,000 images needs 20 GB.
- An 8-megapixel photo viewer with 20,000 images needs 40 GB.
- A 10,000 song MP3 player needs 40 GB.
- A 10,000 song high-definition (HiD) player (like a compressed DVD audio) needs 1.5 TB.
- A 100 movie player at VGA resolution needs 70 GB.
- A 100 movie player at DVD resolution needs >400 GB.
- A combination 20 k 4-Mpixel photo, 10 k MP3 song, and 100 VGA movie player needs 130 GB.
- A combination 20 k 8-Mpixel photo, 10 k HiD song, and 100 DVD movie player needs 1.75 TB.

Based on this table and estimates for the decline in storage costs over the next few years, the storage device (flash memory) for a 20 k 4-Mpixel photo, 10 k MP3 song, and 100 VGA personal video player (PVP) (130 GB) can be purchased for less than $40 in 2017, enabling a consumer product with these characteristics selling for under $400 (depending upon what else is included in the device). With the additional integration of the storage device into the host product suggested in a later section of this book, the net cost would be even less, and a new finished product price of less than $200 should be possible.

Overall storage required for commercial and personal content in the home should explode over the next few years. Personal content should grow even faster than commercial content. This will create a much greater demand for backup and archiving of personal family content either in the home or in the cloud. New technologies must be created to manage this content, organize it, and make sure that it is backed up. Backup could be in the home on external direct attached storage in in-home network storage devices or in a remote cloud storage. At the very least, disaster recovery of personal content could be accomplished using online providers of cloud-based backup services where content in the home is automatically backed up.

1.5 Economics of Consumer Devices

We shall see that consumer devices are very sensitive to price. This creates a focus on reducing the overall product price in order to reach most customers. However, there are lease model approaches for equipment that allows more expensive consumer equipment to be paid off over time (sometimes used for smart phones). We shall also see how the retail channel puts pressure on initial product manufacturing cost since there are usually several layers of markup that also have to be taken into account in arriving at the final product price to the consumer.

1.5.1 Consumer Product Price and Demand

There are significant differences between the retail and service markets for consumer electronic devices. The *retail market* is primarily a push market where retailers advertise products and offer various marketing approaches and discounts to persuade potential customers to make purchases. By contrast, the *service market*, typified by call phone and cable companies, offers the consumer electronics hardware for free, at a discount, or on a low-cost lease basis in order to increase the services purchased as an ongoing subscription by the customer.

The retail market for consumer electronic products is very price sensitive, and digital storage devices are often one of the more expensive components used in these products. Consumer products are often purchased with discretionary funds, so price is a very important factor in a purchase decision. Figure 1.3 represents the willingness of customers to purchase a consumer device as a function of product price.

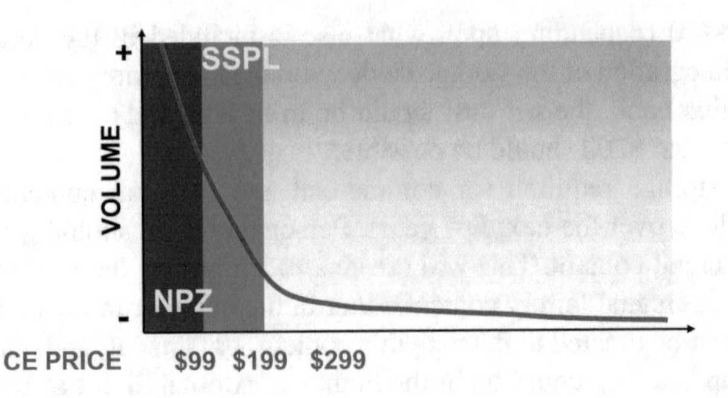

Fig. 1.3 Sales volume as a function of product price (Source: Cornice)

Note that SSPL means *single spouse permission limit* (what an adult could generally purchase without getting into trouble), while NPZ means *no permission zone* (what an individual could probably buy with no negative repercussions).[2] Clearly the sales volume increases significantly as the sales price declines. The SSPL point is estimated to be about between $100 and $200 retail while the NPZ is less than $99. Above $200 the purchase is likely to require at least some family discussion before a purchase is made in order to avoid financial problems or at least some level of heated discussion.

1.5.2 Cost Markups in the Retail Sales Channel

Figure 1.4 shows the markup in price from the initial product cost through the typical retail distribution chain. As can be seen, the amortized effective cost of a digital storage device in the final sale of the consumer device can be more than twice that of the manufacturer's storage product price. With every facility that a product goes through in reaching the consumer, there is additional overhead and handling costs that must be assumed. These must be added to the price that the product sells to the next level in the distribution chain and ultimately to the cost to the consumer. As shown in Fig. 1.3, as the price goes up, the unit volume of sales generally goes down.

The latter consequence is, generally, not always true. If a product is seen as of much higher quality or if it gives the user a greater feeling of status, then they may be willing to pay a higher price than simple market dynamics would indicate. This is the case, for example, with many Apple products. These products are perceived as being "cool" and giving the user a higher status. As a consequence, the price of these products has not declined with time as they would in a normal market. In the long run, unless a company can maintain new technologies that continue to be perceived as "cool," they will eventually lose this status and the price and profit margins will tend to decline.

[2] From 2003 Storage Visions Conference presentations by Cornice Inc.

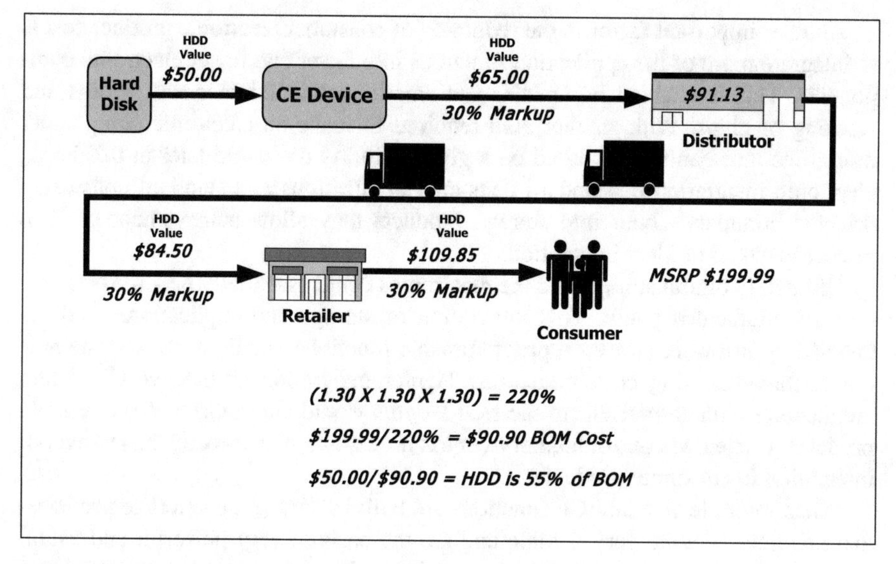

Fig. 1.4 Consumer storage markup through the retail distribution chain

1.5.3 New Opportunities for Electronic Integration

Technologies used in consumer electronic products come from many sources and have varied histories. GPS positioning technologies were originally developed by the military and only in the last 25 years have come into use by consumers. Today GPS-based positioning capability has become one of those standard functions that are being built into consumer electronic products.

Often consumer electronic product technologies are expensive to manufacture at first and then go down in price as product yields increase and unit volumes go up. Optical disk media product introductions often follow this trend. Every new optical disk format from CD to Blu-ray started out costing over $1000 per player and decreased to less than $200 within about 5 years from initial commercial introduction. Consumer markets have market niches, one of which is high-end consumers that, e.g., put elaborate home theaters in their homes. These early adoption consumers drive the initial market for products and often but not always determine whether a product will be successful in a broader market.

There are many sources of consumer product ideas, but whatever the source, they often start out expensive initially and then decrease in price with time. With new technological development, such as less expensive memory, they may also develop more capabilities with time for the same or lower prices. Digital storage is one of the key technologies for many of these applications, and as indicated, for many products, it is one of the largest contributors to the cost of the device. The cost reduction of consumer products is often dependent upon lowering the cost of storage.

Another important factor in the reduction of consumer electronic product cost is to integrate more of the application functions into fewer and fewer electronic components. This is enabled by smaller and smaller semiconductor line widths and stacking of chips. With smaller area required to make an electronic component, more functions can be integrated on a given chip. As discussed later in this book, electronic integration of standard consumer applications and standard consumer-oriented commands built into storage products may allow achievement of even greater levels of product integration.

Ultimately digital storage and the electronics could essentially become one produced manufacturing unit. This integration of storage and applications, perhaps enabled by firmware (software programmable functions), could open a whole new era of manufacturing cost reductions. Tighter integration of the overall system architecture with shorter electronic lead lengths would allow faster, more reliable products. Current visions of memory-centric processing may directly lead to overall integration in consumer products.

When multiple standard CE functions are built into integrated storage products, these devices become very flexible and are the basis of very powerful and useful *convergence devices* (multiple function products). If the firmware for accessing and customizing functions is easy to use and open standard-based, then the time to introduce new products as well as the capability of changing products now in the field could be greatly enhance. In the consumer market, product introduction timing can make all the difference in product profitability. This could have a very great value for the industry.

1.6 Rules for Design of Digital Storage in Consumer Electronics

Based upon observations of what works and what doesn't work in the design of digital storage in consumer electronic products, I would like to propose some rules for digital storage design. These rules, if followed, should help make such products more successful in the market place. These rules are:

- Use the most cost-effective storage component(s) that provide enough performance and capacity for the application.
- Never design a product where you intentionally limit the available storage capacity to the customer—always allow a means of storage capacity expansion.
- If appropriate, incorporate the advantages of multiple types of digital memory to achieve some of their individual advantages. Often a hybrid product using multiple types of storage is better for an application than a single storage device.
- Use electronics and firmware to protect the customers content and battery life.
- Make it easy to back up and copy data (storage is cheap, time is not!).
- Give consumers a way to protect their personal content and privacy (encryption of data on the storage devices could help with this).

- Make storage management and organization automatic—for instance, protect data and prevent replication of corrupt data.
- Design the components including storage to provide lowest total product cost (storage integration concepts could help here).

The basic concept behind these rules is to make the end product more attractive to the consumer both in terms of price, features, and performance. Most of these are common sense observations. The basic idea is for a product designer to be objective; his goal is to provide the customer the best possible product at the best possible price. Providing the best performance at a certain cost may involve using hybrid devices that combine more than one type of storage to get better overall system advantages.

Also in a world where digital content is going to higher and higher resolution, it is good to give the customer access to or at least an easy option to add more storage capacity. This may mean larger onboard storage, but it could also involve the use of wired or wireless storage peripherals that the device can tap into to get reliable access to larger amounts of storage than the device itself can carry. It may also involve access to storage in a public or private cloud.

User content is valuable. In an age where consumers often carry and use content creation devices such as digital still and video cameras in mobile devices that could be subject to loss or theft, protecting this content from unauthorized access is important. Backing it up as soon as possible is a requirement for consumer devices going forward. As consumer-made content increases in volume, creating ways to automate the backup, archiving and protection of this content becomes more critical. Likewise, creating ways to organize consumer content, including *automatic metadata capture* and creation, will make our lives easier as the sheer volume of personal content increase but as our time to manage it, likely, becomes less rather than more. These functions need to be automated, probably in more advanced home storage network systems.

As mentioned before, there will be great advantages available for net product cost, better product performance, and time to market by leveraging electronic integration. This is made possible by denser transistor architectures in storage devices. The result is to build standard consumer electronic application commands into storage device firmware.

1.7 Classification of Devices Using Storage in the Home

We classify five types of digital storage devices or devices heavily dependent upon digital storage used in and around the home, including mobile devices and automotive devices. These are:

1. *Active devices* allowing user interaction with other devices to exchange content files. These include computers and smart phones.
2. *Drone players* that retain content for play out after it has been downloaded from another local source, usually an active device.

3. *Direct attached external storage devices* used to expand local storage of another device or for backup of content on the other device. These external direct attached storage devices may use USB, Firewire (IEEE 1394), eSATA, Thunderbolt, or other less common external interfaces.
4. *Network attached external storage devices* used to provide a central content sharing device or centralized backup of content on other devices.
5. *Static or mobile personal content creation devices* such as a digital still or video camera or perhaps in the future a "life-log" device. Such personal content may then be saved on an active device, a direct attached storage device or a networked storage device.

We will also define consumer devices as *static*, that is, not frequently moved, and *mobile*, that is, often moved. Static consumer devices using digital storage include digital video recorders (DVRs), set-top boxes with DVR capability, home direct attached storage, and home network storage. Mobile consumer devices that use digital storage include many different types of portable music and video players, digital still and video cameras, cell phones, automobile entertainment, and navigation systems.

1.8 Consumer Electronic Storage Hierarchies

We will examine the uses of various digital memories in electronic devices to get an understanding of what sort of trade-offs designers have to make in their electronic designs. Then we will broaden our analysis to look at storage hierarchies for static and mobile consumer applications. We shall use these hierarchies to lay down some general guidelines on what storage to use for different applications, including how different storage devices can be used together to create hybrid storage devices that combine good features of different types of storage.

We can learn a lot about the behavior of a storage system by plotting the touch rate vs. the response time[3]. The response time measures the time to act on a single object, while the touch rate relates to the time to act upon the data set in a storage system as a whole.

Figure 1.5 includes indications of various sorts of applications (performance regions) and general trade-offs important to system design.

The chart shows touch rate as log touch per year on the vertical axis and log response time on the horizontal axis with faster response times on the left. Shorter response time means data can be accessed more quickly, increasing its value. Higher touch rate means more data can be processed in a given time period, increasing the value that can be extracted.

Figure 1.6 shows where various storage technologies lie in the touch rate chart as shown in Fig. 1.5.

Flash, HDD, and tape are used to address workloads in different regions. These storage technology regions have a hockey stick shape due to the way performance limits impact the system performance. On the right side of a region, the performance for small objects is dominated by the access time (time to get to the first byte of data). On the left side, the performance for larger objects is dominated by the data transfer time.

[3] Touch Rate: A metric for analyzing storage system performance, S.R. Hetzler and T.M. Coughlin, 2015.

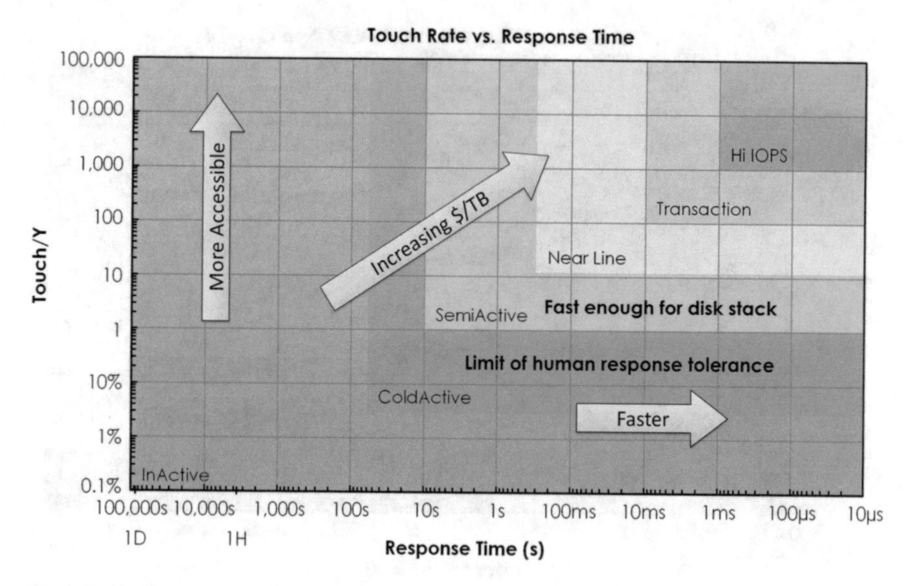

Fig. 1.5 Touch rate versus response time chart

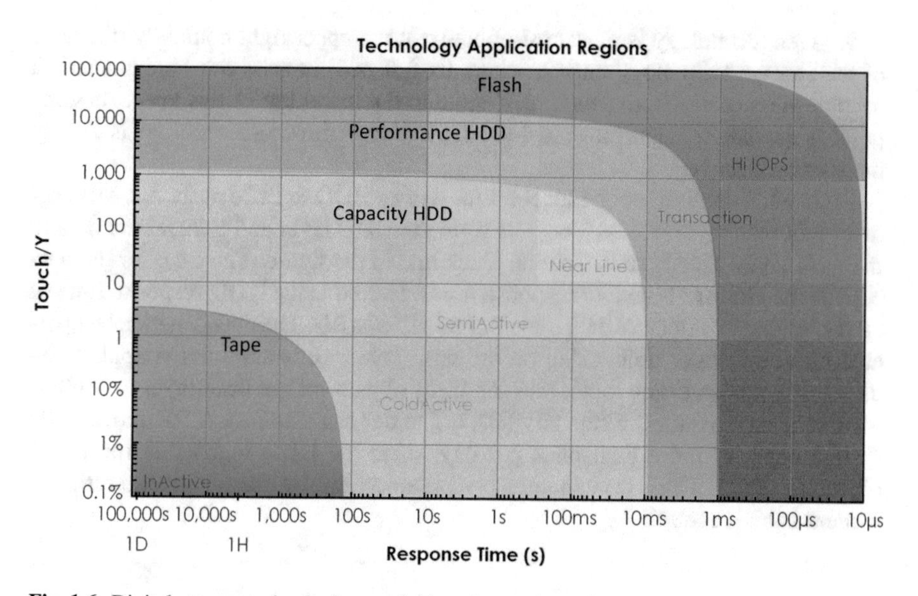

Fig. 1.6 Digital storage technologies overlaid on the touch rate/response time chart

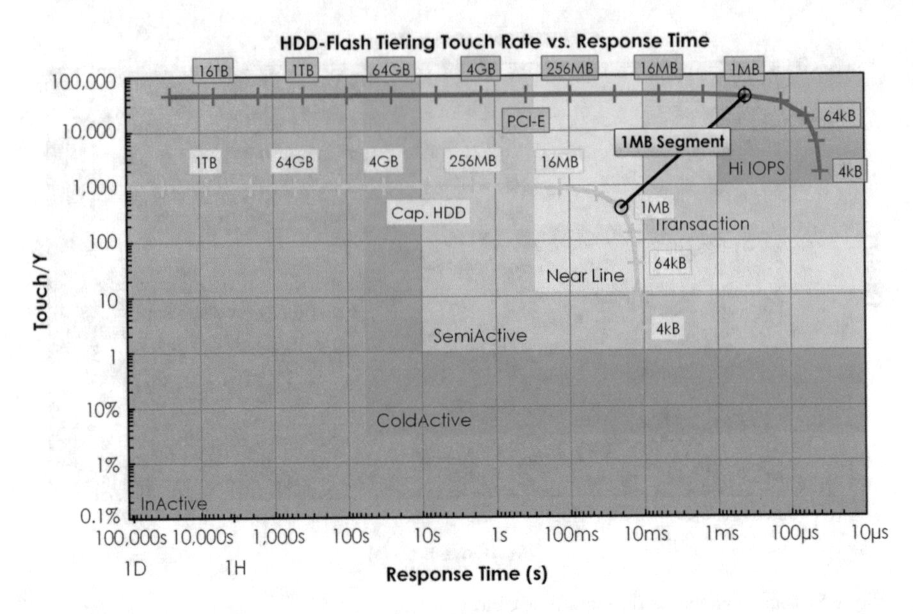

Fig. 1.7 HDD-flash tiering touch rate versus response time

A given technology is most cost-effective at the upper right boundary (the knee of the storage technology curve), where its full performance can be extracted. It becomes less cost-effective as it used more to the lower left of this knee. At some point as the requirements move to the left, a lower performing technology is usually more cost-effective.

Figure 1.7 shows a touch rate chart for tiering HDD to PCIe flash. A 1 MB segment will read random 1 MB objects from the HDD layer and transfer the data to the flash layer. This is shown as the black line in the figure. The time for the data copy is the slower of the two response times. Here it is the HDD response time of 20 ms, giving an effective data transfer rate of 50 MB/s. The data can then be operated on at a smaller object size on the flash layer. Let's assume the application uses a 16kB object. The touch rate for 16kB objects on the flash layer is 6100/Y, while it is only 10 for the capacity HDD. The tier can also access the entire HDD system capacity more than once per day, since the HDD touch rate for 1 MB objects is 400/Y. Thus, we can get a performance improvement if the hit ratio for content is high enough.

1.8.1 *Digital Memory for Device Process Execution*

Memory in digital devices can be used for the execution of microprocessor commands or for longer-term storage. The three types of electronic memory used for process execution are shown in Fig. 1.8. *SRAM* (static random-access memory) is

Fig. 1.8 Speed and price characteristics of common process execution memory devices

the fastest memory but also the most expensive. *DRAM* (dynamic random-access memory) is not as fast as SRAM but less expensive, and it is faster than *NOR* (refers to "neither or" a logical operation) memory. NOR is the least expensive of these execution memories but it is slower. SRAM and DRAM memory technologies are volatile, that is, they lose their memory when it is not refreshed or if power is removed.

SRAM keeps data in the main memory of a processing unit, without frequent refreshing, for as long as power is supplied to the circuit. SRAM is very fast and its access speed is closer than that of the CPU. However, SRAM is much more expensive than DRAM and takes more space than DRAM for the same size of memory. A SRAM bit consists of four to six transistors, which is the reason for the bigger size compared to DRAM. The advantages of SRAM are speed and the lack of a need for constant data refresh. The disadvantages of SRAM are cost and size. SRAM is often used for CPU cache memory (both level 1 and level 2) because of its speed.

DRAM is the most often used RAM in computers. DRAM can hold its data if it is refreshed by a special logic circuit (the refresh circuit). The refreshing reads and rewrites the content of the DRAM memory frequently to prevent loss of the DRAM's contents. Even though DRAM is slower than SRAM and requires the overhead of a refresh circuit, it is still much cheaper and takes about one quarter of the space of SRAM. In a DRAM, a capacitor holds an electrical charge if the cell contains a "1" or no charge if it contains a "0." The refresh circuitry reads the content of each cell (bit) and refreshes each one with an electrical charge before the content is lost.

NOR is a type of flash memory. Flash memory is nonvolatile, which means that it does not need power to maintain the information stored in the chip. Flash memory offers slower read times compared to volatile DRAM memory. Reading from NOR flash is similar to reading from random-access memory, provided the address and data bus are mapped correctly. Because of this, most microprocessors can use NOR flash memory as execute-in-place (XIP) memory, meaning that programs stored in NOR flash can be executed directly without the need to copy them into RAM. The capability of acting as a random-access ROM (read-only memory) allows NOR to provide a low-power, low-cost option to DRAM or SRAM in consumer products. For midrange and low-end mobile phones, NOR flash is quite common.

1.8.2 Mobile Device Consumer Electronic Storage Hierarchy

The prior section discussed memory used in the execution of microprocessor operations. In this and the following section, we will address longer-term data retention with memory that acts as mass storage. Mass storage can include various semiconductor devices as well as magnetic hard disk drives, tape drives, and optical memory. These mass storage products have characteristics that make one a better fit for a particular application than another. Optical drives and tape drives are too large to fit into mobile devices (although there have been designs in the past with very small optical disks or tapes that could and did fit into mobile devices). In Table 1.4, we list some important characteristics of various mass storage devices as well as a relative ranking of these technologies by these criteria where 1 is best and 5 is worst.

These candidate mass storage devices have different attribute advantages. Also, it should be realized that the attributes listed in the table depend on many variables not included in this table. For instance, hard disk drives use less power as the disk gets smaller, so a small hard disk drive may use much less power than a 5.25 inch form factor optical disk drives or most tape drives.

Many of these deficiencies for different storage technologies can be reduced by the use of better electronics and firmware. For instance, the wear that can occur with rewriting of flash memory cells can be reduced with error correction software as well as wear reduction software that spreads writing around so device cells tend to wear more uniformly.

Also, these are comparisons between the actual storage units within the storage device (such as cells in a flash memory). In practice, a hard disk drive could provide faster write speeds than a poorly designed flash memory unit. In any actual design, there are trade-offs taken to create the most optimal solution to match that application's needs.

Computer scientists often refer to the characteristics of various memory devices as constituting a *storage hierarchy*. The concept of a storage hierarchy allows sorting various memory products based on important attributes or characteristics for the applications for which they are to be used. Figure 1.9 gives a typical example of a computer storage hierarchy based upon data access speed. In this figure, cache memory is fast SRAM, while the main memory for the CPU (central processor unit) or microprocessor is DRAM. Hard disk drives are slower than DRAM but faster than the removable devices such as optical disks or USB drives.

Below is a list of important characteristics for consumer electronic storage devices that we will use to construct a mobile consumer electronic storage hierarchy. Mobile consumer electronic products include music and video players, cell phones, and many other products.

- Price
- Size
- Power
- Capacity
- Data rate
- Reliability and environment

Table 1.4 Ranking of various attributes for common consumer electronic mass storage technologies

Memory	Write speed (storage unit)	Read speed (storage unit)	Rewritability (storage unit)	Environmental sensitivity (shock, vibration, humidity)	Access speed (time to get to data location)	Power used	$/GB	Base cost $	Comments
NOR flash	3	1	5	1	1	1	5	1	Lowest base cost
NAND flash	2	2	4	2	2	2	4	2	Low base cost, fast read
Hard disk drive	1	3	1	5	3	3	3	4	Lowest $/GB for internal storage device
Optical disk	5	5	3	3	5	4	2	3	Lowest cost for content distribution
Magnetic tape	4	4	2	4	4	5	1	5	Lowest media cost for rewriting

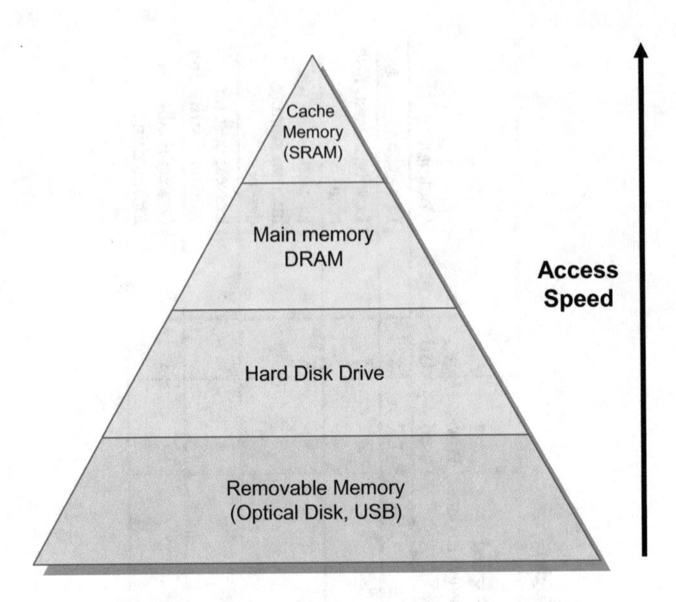

Fig. 1.9 Traditional computer storage hierarchy

Table 1.5 Comparison of mobile storage product important characteristics

	SFF HDD (2.5 inch)	MLC SD flash (2.5 inch)
OEM price	\$50–\$90	\$50–\$130
Capacity	1–2 TB	128 GB–256 GB
Height	7.0 mm	7.0 mm
Width	69.85 mm	69.85 mm
Length	100 mm	100 mm
Write data rate	144 MB/s	470 MB/s
Read data rate	144 MB/s	540 MB/s
Active power	1500 mW	3400 mW
Standby power	100 mW	70 mW
Temperature range	0–60 C	0–70 C
Operating shock	400 G	>1500 G
Nonoperating shock	1500 G	>2000 G
Noise	Some	None
Washer resistance	Low	High

The storage products that we wish to consider in this hierarchy are SFF hard disk drives and NAND flash memory. Table 1.5 compares these characteristics for these mobile storage devices at the time of writing. Storage capacities in this table are not representative of all products and reflect the situation in 2017. Note that the last row, washer resistance, refers to the survivability of a removable storage device to a clothes washer cycle. Note that a magnetic tape is used in some digital video cameras and could also be included in a mobile hierarchy.

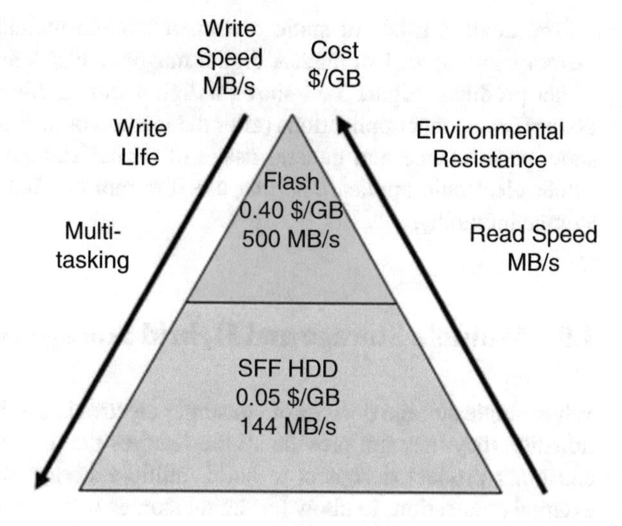

Fig. 1.10 Mobile consumer electronic storage hierarchy

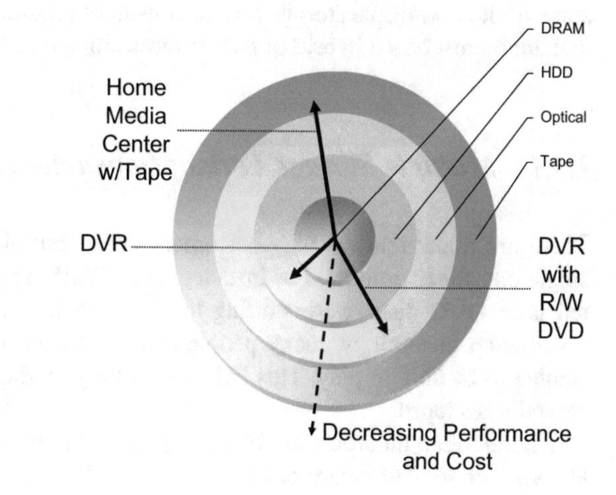

Fig. 1.11 Static consumer electronic storage hierarchy showing a performance or data access speed hierarchy. Arrows are also shown in this figure representing the storage hierarchy elements used in some common static consumer devices

In Fig. 1.10, we make an attempt to construct a mobile consumer electronic storage hierarchy. In this figure, we include NAND flash memory and small form factor hard disk drives.

1.8.3 Static Device Consumer Electronic Hierarchy

For static or fixed consumer applications such as *digital video recorders* (DVRs) or *home media servers*, the important considerations for the choice of digital storage include storage capacity and price, read and write speed, ability to handle multiple content streams, vibration, acoustical noise, and long-term reliability.

The devices used for static consumer storage include hard disk drives, flash memory, and optical storage. A digital magnetic tape was used in some older consumer products. Figure 1.11 shows a digital storage hierarchy for static (e.g., stationary) consumer applications (after the method of S. R. Hetzler[4]). In this view, we show performance and general usage of digital storage devices for several consumer electronic applications that use different combinations of elements of the storage hierarchy.

1.9 Multiple Storage and Hybrid Storage Devices

While single storage devices of consumer electronic products may give basic functionality, they may not provide all the features that customers want. This can lead consumer product designers to build multiple storage devices into the device or external connections to allow for digital storage expansion. This section will look at some of these multiple storage format consumer products as well as storage devices that are themselves a hybrid of two or more storage technologies.

1.9.1 Multiple Storage Format Consumer Devices

There are many reasons why customers want multiple storage technologies in a single consumer product. For instance, when VHS tapes were phasing out and the red laser DVD disks were starting to take over as the primary customer content distribution technology, many people had collections of VHS tapes that they still wanted to be able to play. This led to a thriving market for DVD players that also played VHS tapes.

Digital video recorders are often shipped with only enough storage capacity for 80 hours or so of standard definition content (for HD content, this would be about 35 hours). After recording for a few weeks, customers could easily fill the hard drive of a DVR and be forced to erase older content to make way for new content. If the customer didn't want to erase the old content, he/she didn't have many options. This problem has led to a couple of solutions, one short term and the other long term. Both use multiple storage technologies to solve the problem of limited internal DVR storage capacity.

A very popular category of DVR products, especially in Asia, has been a combination DVR and optical disk recorder that can record content from the DVR onto optical disks so the content can be saved even if the HDD in the DVR is erased. Figure 1.12 shows an example of a DVR with DVD recorder.

[4] S. R. Hetzler, The Evolving Storage Hierarchy, Presented at the INSIC Alternative Storage Technologies Symposium, Monterey, CA 2005.

Fig. 1.12 DVR with DVD recorder for saving old content before the HDD is erased

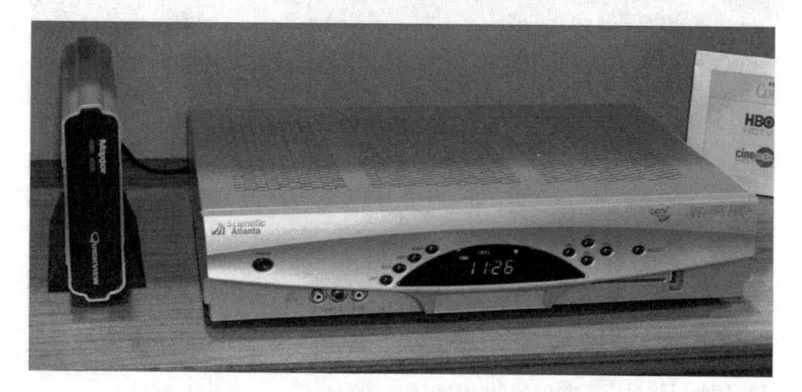

Fig. 1.13 eSATA storage expansion box attached to an eSATA interface on a digital video recorder enabled set-top box

The other approach to expanding the storage on a DVR is to allow an external storage device to be attached to it that can expand the effective storage capacity of the DVR without replacing the internal hard disk drive. Many set-top and even stand-alone DVRs come with an *external SATA* (eSATA) port on the back. eSATA allows data to be transported over the connector cables up to 6 Gbps (750 MBps). Such very fast connections allow fast transport of HD and UHD video content.

The eSATA port allows connecting an eSATA external storage device to the DVR. For instance, a 4 TB (4000 GB) hard drive external storage device with an eSATA interface could give significant additional storage capacity to a DVR with a 500 GB internal hard drive. Figure 1.13 shows an external eSATA storage device attached to a set-top box.

These are only a few examples of combining multiple storage technologies together to create a more attractive product to the consumer. Such enhancements also have the beneficial effect for the consumer electronics company of allowing the sale of these combination products at a premium over and above the cost of adding the extra features. This helps create differentiation and gives customers extra features that they really want.

There are many other examples of combined storage products in consumer devices. Computers include a combination of many different storage technologies. They may have a hard disk drive, an optical drive, multiple USB connections, and often places to directly plug in flash memory cards—usually to transfer photographs to the PC. This approach is a common one, allowing customers to make their choices of what they prefer and to allow for storage technology transitions.

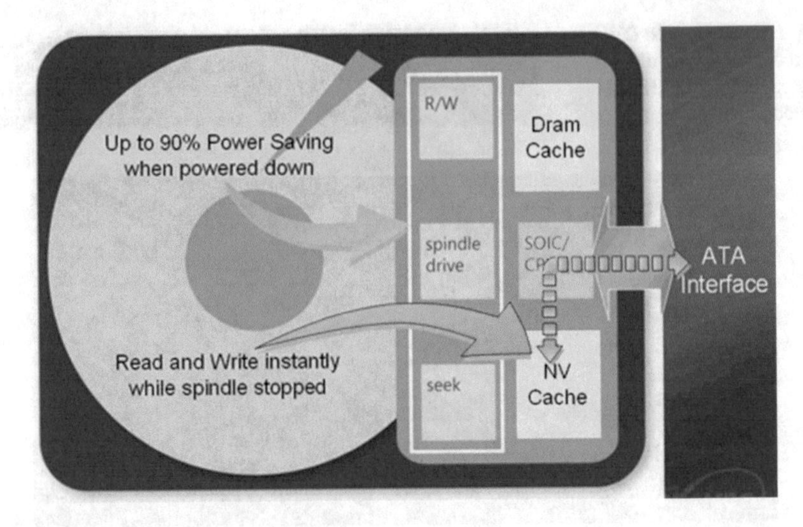

Fig. 1.14 Hybrid hard disk drive architecture and operation

When USB drives first came out, computers for many years came with both USB connections and floppy disk drives. Today floppy disk drives are only available as USB external products to allow reading older content. Note that the move to using wireless technologies has changed some of this need for multiple storage technologies, and some companies, such as Apple, have eliminated additional ports into their computers.

1.9.2 Hybrid Storage Devices

In a way, a storage device that combines a solid-state storage technology with another type of storage is a step toward application integration into the storage product. For instance, *hybrid hard disk drives* (called solid-state hybrid drives by Seagate) add a large NAND flash memory cache on the circuit board of a hard disk drive. This NAND cache can be used to save content that can be written to the disk drive later or for caching frequently used content for rapid access.

Figure 1.14 shows a hybrid HDD drive architecture. This technology was first being introduced into 2.5 inch hard disk drives for laptop computers but is now available for 3.5 inch HDDs as well. The disk drive controller is set up to handle the conventional volatile DRAM buffer found in hard disk drives as well as a nonvolatile flash memory.

Write data is cached into the nonvolatile flash memory, and the disk drive is left in a low-power mode until it is needed either to read data that it contains or when the nonvolatile flash cache becomes nearly full and its contents must be flushed to the hard disk drive to make room for more cached data. Frequently accessed data can also be stored on part of the flash memory as a read cache. Generally as the flash cache gets filled, old data is erased (after being transferred to the HDD, if needed).

Data can be read from the disk via the DRAM or directly from the nonvolatile cache memory. In addition to this caching function, some of the flash cache can be "pinned" that is kept in the cache. This would likely include some of the boot up data for the operating system. Since the flash drive can be ready to transfer data faster than the hard disk drive when the drive is powered up, the boot of the operating system can begin with the "pinned" data in the cache.

A hybrid disk drive could leave the hard disk drive in a low-power mode for a considerable time before the cache needs to be emptied to the hard drive or until data must be read from the hard drive. As a consequence, a hybrid hard disk drive can save a considerable amount of power when compared to a non-hybrid hard drive. Even with a 50% decrease in the hard disk drive consumed power, the total power savings in the laptop may only be about 12% because other components such as the display consume so much power. In addition, it can boot up a bit faster than the conventional laptop hard drive. In practice, the hybrid drive shaves a few seconds off a normal boot time.

There are also *solid-state drives* (SSDs) available for the mobile notebook market which use lower power than hard disk drives (although again the impact on total system power is minimized since storage is often a small part of total laptop power). SSDs also provide faster boot times. However, to have a storage capacity that approaches that of 2.5 inch hard disk drives (over 1 TB when this was written), the laptop computer would cost over $200 more. A hybrid hard drive also has higher shock and vibration endurance since the drive disk is not spinning with the head loaded most of the time.

Thus, a hybrid storage product that combines good features from multiple storage technologies may represent an interesting architecture for consumer design, particularly mobile devices. Unfortunately, hybrid HDDs have never really taken off, and with the declining price of SSDs, they are gradually displacing HDDs, particularly for mobile computer applications.

SSDs may have their own form of hybrid storage by combining various types of flash memory in the same box. For instance, *single-level cell* (SLC) and *multiple-level cell* (MLC) NAND flash memory can be combined together in a hybrid device. This hybrid device uses lower-cost MLC storage for data that is frequently read, while the SLC memory is used for data that is often written. Since MLC flash sacrifices the number of possible erase cycles it can go through to multiply the flash cell storage capacity, this could give a larger storage capacity for a given price while helping with the long-term reliability of the overall device. The same sort of approach can be taken for some 3D MLC combined with bulk TLC or QLC storage.

Hybridization of storage technologies can be a great way to leverage the advantages of the individual storage technologies. Successful hybridization requires good knowledge of the specifications, performance of the individual storage technologies, as well as possible interactions that the two technologies can have together. Hybrid storage is definitely a tool that consumer product designers want to keep handy in their design tool box.

1.10 Chapter Summary

- Large digital storage capacities at an affordable price are key drivers of new consumer electronic products. With larger storage capacities, consumers can store and access even larger collections of higher-resolution content. High-end audio/video players require several hundred GB of storage capacity.
- The consumer electronic market is a very price-sensitive market. Products with prices greater than about $200 generally sell in lower volume than those less than $200, and the volume increases even further when the price drops below $100.
- In many consumer products, the digital storage is a large part of the total system cost. The initial cost of the storage to the consumer electronic company is magnified by the overhead from the distribution chain to get to retail stores.
- With greater levels of electronic integration possible and with many consumer electronic functions become standardized, it should be possible to mate consumer applications and storage electronics to result in lower total product cost, greater reliability, and faster performance.
- A number of common sense rules have been developed for the design of digital storage in various consumer products.
- We reviewed memory options and characteristics for cache and main memory of microprocessors.
- Digital storage devices have different advantages depending upon the application. Since static and mobile consumer devices have some common characteristics, we developed digital storage hierarchies that can be used to choose the proper digital storage for the application.
- Different storage technologies can be combined together in consumer products to meet technology transitions and to enhance a product so it is more appealing to the customer and to enhance the profitability of the consumer electronic company.
- If one storage technology is directly combined into another storage product, we can produce a hybrid storage device. Hybrid hard disk drives combining flash memory as a cache on a hard disk drive are an example of this. Such hybrid products can provide some of the advantages of the individual storage technologies and should be carefully considered by consumer electronic design engineers.

Chapter 2
Fundamentals of Hard Disk Drives

2.1 Objectives in This Chapter

- Learn about the history and uses of hard disk drives.
- Describe the basic layout and operation of hard disk drives.
- Clarify the organization of data on a hard disk drive.
- Understand the self-monitoring and reliability specifications for hard disk drives.
- See how disk drives are designed for various consumer applications.
- Enumerate the factors that determine the cost of a hard disk drive and its impact on CE product price.
- Go over the developments of hard disk drive electrical interfaces.
- Look at the future development of hard disk drive technology and the speed of areal density growth.

2.2 History of Hard Disk Drives

Hard disk drives have a long history. Magnetic recording was first patented by a Danish inventor, Vladimer Paulsen, in 1898 using electromagnets and magnetic wires. Up to the 1950s, engineers developed ways to use magnetic recording to record analog music and other recordings on various media including tapes and disks. In 1956 IBM introduced the first digital magnetic disk drive, the RAMAC (random-access method of accounting and control). The RAMAC had a total storage capacity of five million characters (3.75 MB) on 50 24-inch disks in a device the size of a large washing machine. It was leased annually to companies for several thousand dollars. Figure 2.1 shows the 1956 IBM RAMAC computer and storage system.

Since 1956 hard disk drives have increased in capacity while decreasing in size and cost. In 2017 10 terabytes of data could be purchased on a single 3.5-inch (95 mm) disk drive for about $350. For many years hard disk drives increased at

T.M. Coughlin, *Digital Storage in Consumer Electronics*,
https://doi.org/10.1007/978-3-319-69907-3_2

Fig. 2.1 The RAMAC computer with the first digital hard disk drive system

about 60% and even over 100% annually (in the 2000s) in storage areal density (the amount of digital storage capacity per usable surface area of the magnetic disk). The 60% annual increase resulted in a doubling of hard disk drive capacity about every 18 months. This rate has slowed in recent years resulting in slower annual changes in hard disk drive available storage capacity. Smaller disk drives can increase in storage capacity faster than larger disk drives because the smaller disks have lower data rates, making design at higher densities easier.

Hard disk drives are used for a great many applications including high-performance 15,000 RPM drives used in enterprise applications and high-performance storage arrays. They are also used in nearline low-cost high capacity storage arrays, desktop computers, and laptop computers. The greatest interest to our readers is that they are used in a number of static (mostly nonmobile) and mobile consumer applications. Hard disk drives are often the "mass storage" for the application that they are used in since they provide a large amount of nonvolatile rewritable digital storage at a lower cost than all other storage devices except magnetic tape. A list of hard disk drive producers is given in Appendix C.

Figure 2.2 shows historical shipments from 2000 to 2016 and projections for hard disk drives for various applications out to 2021, while Fig. 2.3 presents projections for hard disk drive form factors through the same period.

Consumer electronics, external storage devices, and, to a lesser extent as time goes on, mobile computers, will be an important contributor to hard disk drive shipped volumes. While 2.5-inch hard disk drives had been increasing as a percentage of total HDDs up to 2013, it looks like 3.5-inch drives will predominate in the future as nearline and external storage applications dominate.

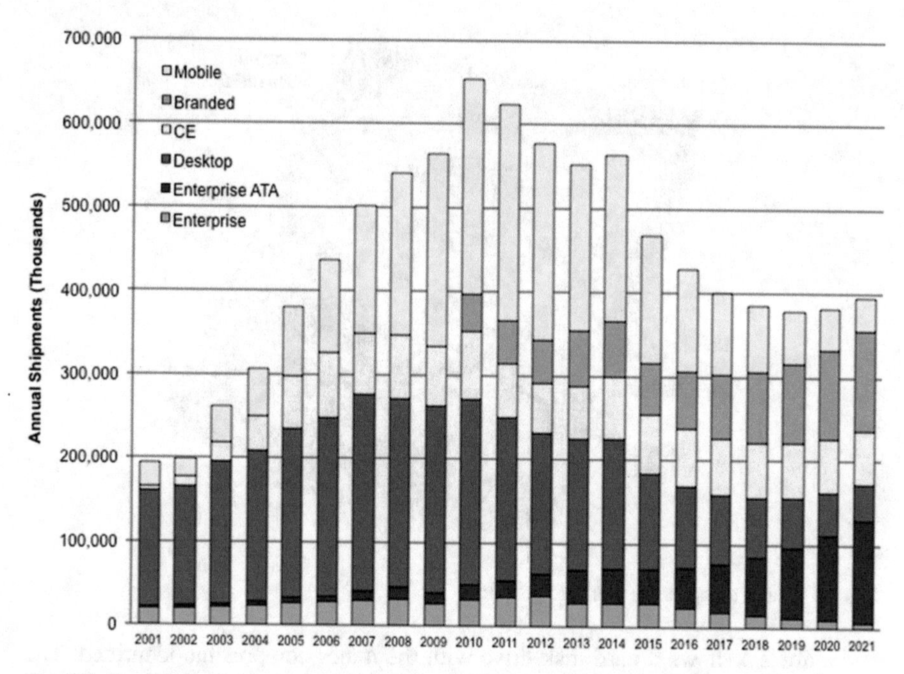

Fig. 2.2 Hard disk drive market niche projections to 2021 (2016, Coughlin Associates)

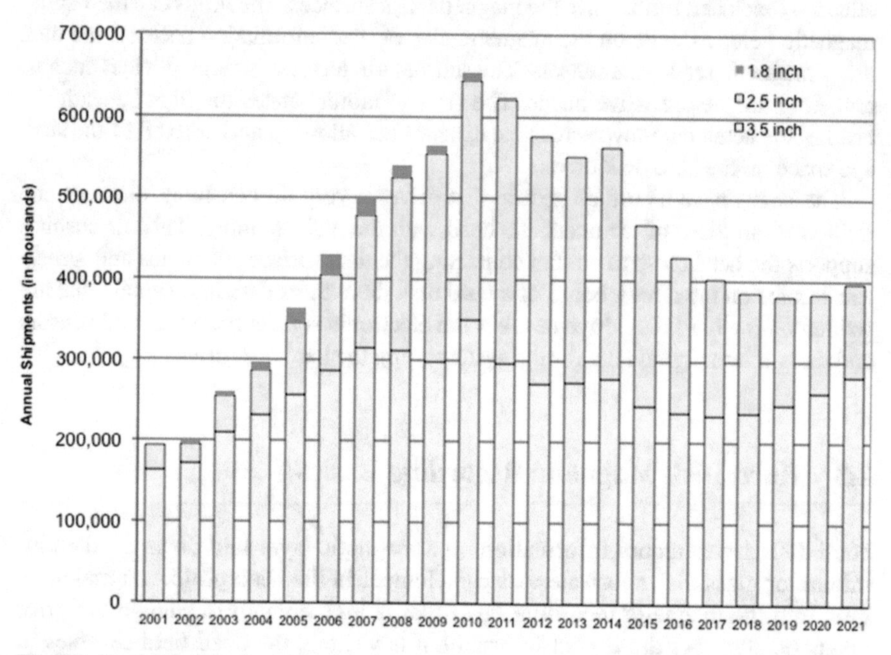

Fig. 2.3 Hard disk drive form factor projections to 2021 (2016, Coughlin Associates)

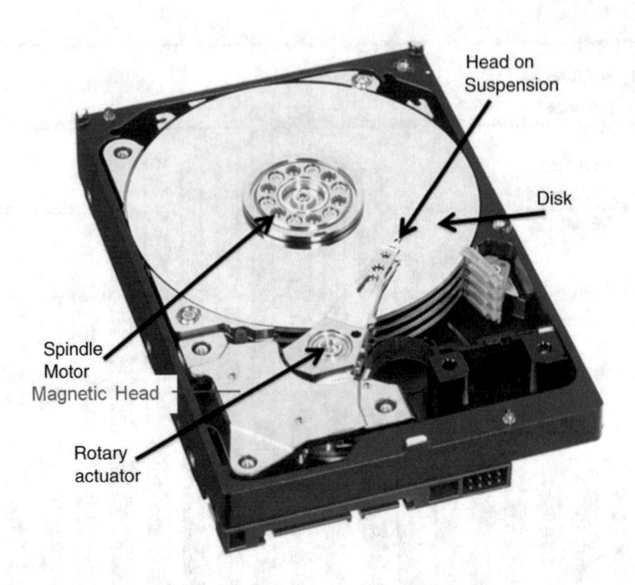

Fig. 2.4 Major components of a hard disk drive

Figure 2.4 shows a hard disk drive with the major components identified. The actuator motor moves the suspensions (at the ends of which the magnetic heads are attached) back and forth above the magnetic disk surfaces. The heads can then apply magnetic fields to write on the magnetic disk or read information back off the disk using magnetic read-back sensors. The sensors use nanotechnology devices such as tunneling magnetoresistive heads. The spindle motor rotates the disk beneath the head as the actuators move across the disk surface allowing access to all of the storage space on the hard disk drive.

The heads have a surface texture that drives a very thin cushion (close to one million of an inch) of air under the head with the disk spinning. This air cushion supports the head close to but not contacting the disk surface as it reads and writes. Electronics on the circuit board of a hard disk drive turns data into signals that are written on the hard disk drive or takes the electronic signals from the read sensors and turns it into digital data for the system using the hard disk drive.

2.3 Hard Disk Magnetic Recording Basics

Hard disk drives record information on a magnetic layer laid down on the aluminum or glass disk substrate. A write element in the head projects a magnetic field into the magnetic recording layer that is high enough to change the prior magnetic state. As the digital information is written, the head field reverses in orientation, writing magnetic transitions between magnetized areas with opposite magnetic orientation. The oppositely orientated magnetic regions in the recording layer create fields above the disk surface that are detected by the read sensor in the head.

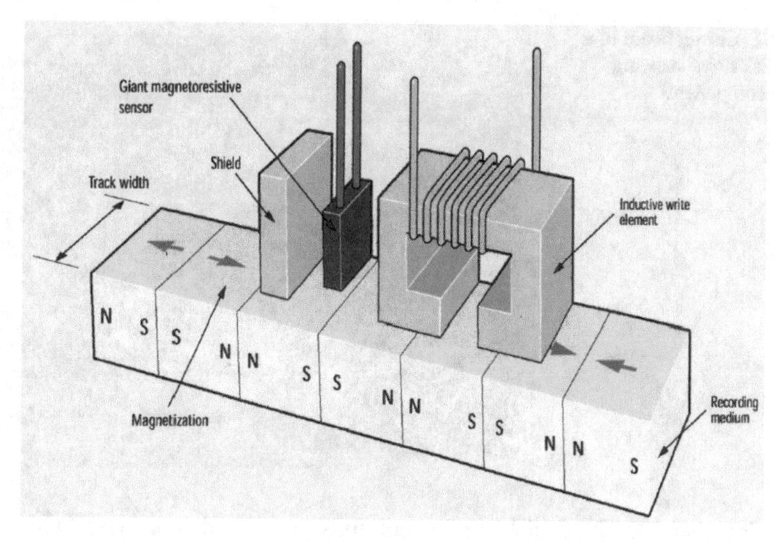

Fig. 2.5 Longitudinal magnetic recording schematic

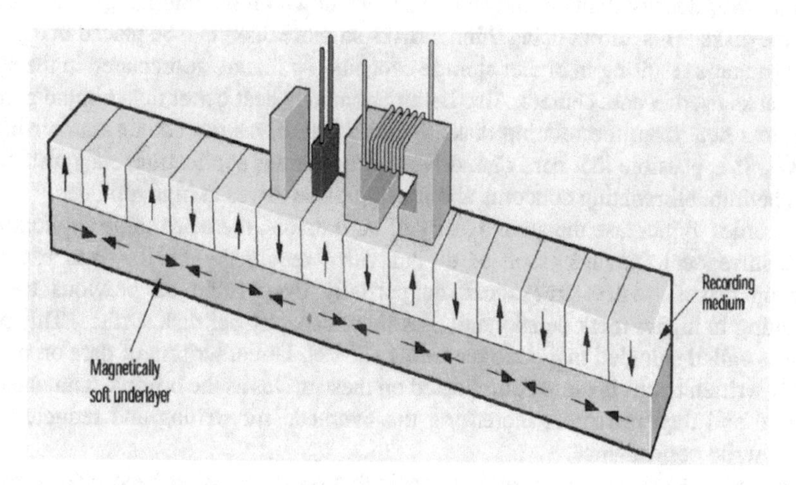

Fig. 2.6 Perpendicular magnetic recording schematic

As the recorded disk spins under the read head, the signals from the read sensors are decoded by the drive electronics to yield digital ones and zeros. A hard disk drive is generally designed to have a useful life of at least 5 years, although the actual drive specifications may differ depending upon the intended application and usage environment.

From the invention of the digital hard disk drive in 1956 until 2005, the recorded magnetization in the magnetic recording layer was always oriented along the plane of the disk in commercial disk drives (this is called *longitudinal recording*). In 2005 Toshiba and Seagate introduced the first hard disk drives using *perpendicular recording* where the recorded magnetizations in the magnetic recording layer are normal to the disk surface. Figures 2.5 and 2.6 are schematics of longitudinal and perpendicular magnetic recording, respectively.

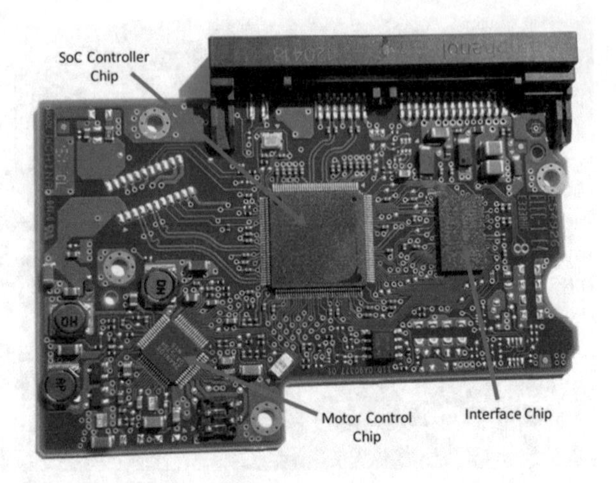

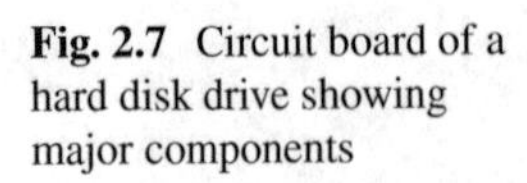
Fig. 2.7 Circuit board of a hard disk drive showing major components

Today there are hard disk drives, currently used in data centers, which are hermetically sealed from the outside environment and are filled with helium. Helium has a lower density than air and so its use cuts down on the internal gas resistance for the disks. This allows using thinner disks so more disks can be placed in a given size package resulting in higher storage capacity—a feature appreciated in the storage racks used in data centers. Also helium conducts heat better than air and generates less heat from the rotating disks, hence these drives run cooler than air-filled drives. It is possible that hard disk drives for consumer applications may also contain helium, eliminating concerns about using these drives at high altitudes.

In order to increase the areal density of hard drives, there are some applications with infrequent rewrites (such as digital video recorders—DVRs—and external backup drives), where tracks can be partially overwritten on previous tracks, resulting in higher track density and thus more capacity per disk surface. This process is called shingled magnetic recording (SMR). Direct writing of data on previously written tracks is more complicated on these drives as the original data must be moved and then rewritten, increasing the overhead for writing and reducing the HDD write performance.

A hard disk drive has a pre-amplifier chip that is located on the head suspension or on the arm near to where the head suspension is attached. The pre-amplifier increases the voltage amplitude from the read sensor in the head, which is then transmitted by wires to the *printed circuit assembly* (PCA) that is attached to the disk drive. Figure 2.7 shows the circuit board in a hard disk drive with key components labeled.

Fifteen years ago, a disk drive circuit board was a large complex board that occupied one entire side of a hard disk drive. Separate ICs were used for all the drive controls and the read/write channel.

With the shrinkage of electronic circuit dimensions, it has become possible to incorporate more functions into a given IC. This has reduced the chip count on a hard disk drive PCA and also reduced the number of discrete components that were once used to tune the circuits in a disk drive.

Today *System-on-Chip* (SoC) designs for hard disk drives can incorporate many functions into one chip that formerly were done with separate ICs. As a consequence, modern hard disk drive printed circuit assemblies (PCAs) are much smaller, typically occupying only a fraction of one side of a hard disk drive.

Greater electronic integration has reduced the cost of hard disk drive PCAs, increased their reliability as well as durability and allowed more functions to be incorporated in a hard disk drive than were possible in the past. Greater electronic integration also made it possible to make hard disk drives as small as 1-inch and 0.85-inch form factors.

Electronic integration in hard disk drives continues. It is likely that hard disk drive boards will shrink further. Eventually we could have single chip designs with significant on-board intelligence. This could include built-in commands that can be used to create applications, lower production cost, and provide greater reliability.

Greater intelligence built into hard disk drives is allowing them to function with lower power, generate less heat and noise, anticipate user actions, and provide advanced caching for writing and reading.

2.4 How Data Is Organized on a Hard Disk Drive

A hard disk drive is organized so that most of its surface is used for user data. There is also some surface area of the disks that are used to store data used internally by the hard disk drive. The data on a hard disk drive consists of bits recorded along tracks on the disk surface. The total capacity of the hard disk drive is the product of the number of tracks times the average number of bits along those tracks.

The *tracks* are recorded in concentric circles around the disk surface. These concentric tracks are broken into *sectors* of a fixed number of bits. Servo information is written on the tracks during the manufacture of the disk drive in recorded servo patterns at the beginning of sectors that give different signals as the head goes off-track in one direction or another. The head detects the servo signal as the head actuator moves the head across the disk surface. The resulting signal is used to position the head over the track using a feedback circuit tied into the head actuator drive electronics.

The sectors have driven accessible data recorded at their start, known as sector headers. The sector headers are used to identify the user data that follows and also provide parity data for error correction. Figure 2.8 illustrates the organization of data on a disk surface.

The size of the disk sectors for many disk drives was fixed to 512 bytes for many years because of the requirements of popular computer operating systems, but changes in the popular operating systems have enabled using sector sizes as large as 4096 bytes. The sector header information at the start of the sector must be read before the user data can be read. If the sectors are smaller, the sectors are located closer together and the ratio of user data to drive data is lower, adding overhead to the percentage of the raw disk capacity that can be used for user data. By using larger sector sizes, higher-capacity hard disk drives can be created, and sustained

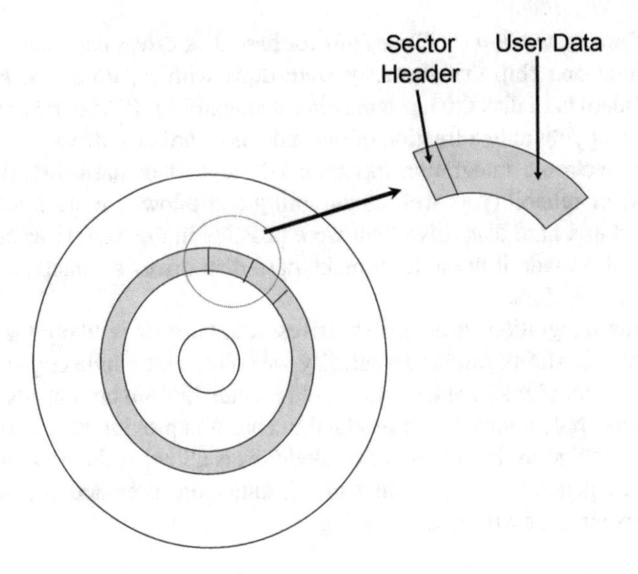

Fig. 2.8 Data is organized on a disk into concentric tracks that are broken up into data sectors

data rates off of the hard drive can be increased. Larger sectors make a lot of sense for consumer applications where the data may be streamed off of a hard disk drive.

Because the linear speed of disk tracks decreases going from the outer radius to the inner radius, in order to maximize storage capacity, disk drives use zoning to change the data rate to and from the disk during writing and reading as a function of the track radius. This maximizes the recorded storage capacity on the disk. Modern disk drive circuits are also capable of optimizing the signal and recording density to accommodate specific capabilities of the actual heads and disks in the drive. Thus, an optimum head and disk combination will record higher recording density than a lower performing head and disk combination.

The advertised hard disk capacity is the "formatted" disk capacity that is accessible by the user. The actual raw storage capacity of a disk drive is higher since the drive can contain varying amounts of nonuser accessible data at the sector headers and in nonuser accessible tracks. The hard disk drive also contains spare tracks and sectors that can be brought into use if an existing sector develops hard defects. Disk sectors containing hard defects that have been retired by the hard disk drive control circuitry are referred to as "g" list sectors (for sectors with grown defects).

Most modern hard disk drives only allow access by users to the logical data structure on the disk and not the actual physical layout of the data on the disk. The mapping of logical data to the physical data is conducted internally by the hard disk drive and is transparent to the user. The development of virtualization of physical data to logical data has made it possible to create disk drives that are self-configurable and requires much less work by user in setting them up. This internal intelligence in hard drives was made possible due to the decreasing line-width of semiconductor processes. Narrower line-width circuitry allows more processing power and memory in modern hard disk drives (at the same time decreasing the actual size of the hard disk drive circuit board).

Table 2.1 List of SMART variables monitored by hard disk drives

Attribute #	Attribute	Units
1	Raw read error rate	Read errors
2	Throughput performance	?
3	Spin up time	s or ms
4	Start/stop count	Spindle start stop counts
5	Reallocated sectors count	Number of uncorrectable errors in sector
6	Read channel margin	Function not specified
7	Seek error rate	?
8	Seek time performance	Ave performance
9	Power-on hours	Hours, minute, seconds
10	Spin retry count	Only if first spin unsuccessful
11	Recalibration retries	Only if first attempt unsuccessful
12	Device power cycle count	Count of power on/off cycles
13	Soft read error rate	Program read errors
191	Gsense error rate	?
192	Power on retract count	?
193	Load/unload cycle count	Load/unload or landing zone cycles
194	Temperature	Degrees centegrade
195	HW ECC recover	?
196	Reallocation event count	# Attempts to transfer to spare sector
197	Current pending sector count	# Sectors waiting remap
198	Uncorrectable sector count	Number of uncorrectable errors in sector
199	UltraDMA CRC error count	Errors in data transfer
200	Write error rate	Errors while writing sector
201	Soft read error rate	Like #13?
220	Disk shift	Unknown measurement of shift
221	G-Sense error rate	Indication of impact loads

Modern hard disk drives also contain capabilities for internal monitoring of important variables that can affect drive performance and reliability such as temperature and various types of defect data. Table 2.1 shows a list of common *Self-Monitoring, Analysis and Reporting Technology (SMART*) variables that hard disk drives support. Note that the SMART commands and monitoring variables are part of the ATA drive specification. We will discuss the ATA specification later in this chapter.

Most hard disk drives also have built-in commands that will make the disk drive erase all the user data on the drive. This command, called *secure erase*, causes the drive to overwrite all the data on the drive so that it cannot be recovered. The "secure erase" still leaves the nonuser data intact so that the disk drive can be re-used. Note that the SMART commands and monitored variables are part of the ATA specification.

Almost all hard disk drives are internally encrypted with an encryption key that never leaves the HDD. This is done using encryption technology built into the HDD electronics. The Storage Work Group of the Trusted Computing Group champions this type of encryption. Some of these HDDs make the encrypted function available to the user so they can protect their data using a password that allows access to the unencrypted data using the internal encryption key. If the HDD is told to erase the existing encryption key (and create a new one), this is called crypto-erase. When the encryption key is erased, the original data cannot be recovered. Since encryption key erasure is very fast, crypto-erase is a lot faster than the secure erase discussed earlier.

2.5 Hard Disk Drive Performance and Reliability

Hard disk drives access data by moving the heads across the tracks of the disk and by spinning the disk under the heads to find the location of a desired block of data. The disk drive keeps track of the location of data and how that data is connected via a host computer-based file system. The data rate of hard disk drives depends upon the speed of rotation of the disk, the diameter of the disk, and the speed and accuracy of the head moving across the disk surface. For a 3.5-inch (95 mm) disk drive used in consumer applications such as digital video recorders, the sustained data rate of the hard disk drive is usually over 50 MB/s.

The sustained data rate is the rate for which the data from the disk drive can be continuously streamed off of the disk drive. Video content with standard definition (SD) has a data rate of about 6 MB/s while high definition (HD) content data rate is >20 MB/s. Ultrahigh definition (UHD) content requires data rates of >30 MB/s. Since the data rate of the hard disk drive is greater than that of individual SD or HD streams, the disk drive can support multiple streams. For instance, the hard disk drive can buffer up content for one stream that it is playing out. While that data plays out from the read buffer, the controller logic takes data being recorded from a write buffer and records it on the hard disk drive. In principal with 50 MB/s sustained data rate, 9 independent SD streams, 2 HD streams, or 1 UHD stream can be supported. In actual application, the overhead in changing from read to write will reduce the available streaming bandwidth and thus the number of streams that can be supported using a single disk drive.

Hard disk drives are rated for their warranted lifetime and estimated time to failure. A typical expected useful life for a hard disk drive is about 5 years, and warrantee periods vary from 1 to 5 years. The mean time to failure (or *mean time between failures*, MTBF) for hard disk drives used for consumer applications is generally 500,000–1000,000 h. Factors that can accelerate hard drive failure include:

- Start/stop cycles—number of disk motor on/off cycles
- Read duty cycle—% of time drive reading during read stream
- Write duty cycle—% of time drive writing during write stream
- Ambient temperature
- Other environmental conditions such as vibration and shock

The number of start/stop cycles is influenced by the number of user on/off cycles as well as power settings for a disk drive that may turn various elements on and off as required. The number of read/write duty cycles is influenced by the size of read and write buffers, the required data rate to support continuous play, and the expected write/read usage depending upon the application. As an example, a digital video recorder (DVR) (also known as a personal video recorder (PVR) typically does a lot of writing of new recorded content with somewhat fewer reads).

The ambient temperature of the drive is controlled by heat dissipation in the host device by way of active cooling devices (fans) and passive cooling devices, the use and frequency of power saving modes, drive duty cycles, and number of writes and reads. By monitoring the SMART log entries for internal drive temperature, usage may be adjusted to keep the internal temperature within acceptable limits.

Shock is especially important for mobile applications. Vibration can be an issue for both mobile and static consumer applications. Moisture and humidity are generally bad for electronic products, including hard disk drives. Temperature cycling in mobile and automotive applications can be an issue in drive performance and reliability since thermal expansion can cause off-track issues, and condensation can cause reliability issues. It should be noted that more than half of all hard disk drive failures are due to circuit board or firmware failures, while the balance of the failures are due to a variety of other causes. This is why data on a disk drive can often be recovered using a different circuit board or with a firmware update.

Note that the sensitivity to mechanical issues such as shock and vibration generally are less for smaller form factor hard drives since the inertia of the disks is reduced with smaller form factors. Also, the total storage capacity, which is the product of the surface area and the areal storage density, is less for smaller form factor disks than for larger form factor disks. Heat generated by a hard disk drive and the power required to run it decreases as the disk radius gets smaller. In addition, smaller form factor hard drives are often quieter.

2.6 Hard Disk Drive Design for Mobile and Static CE Applications

Mobile devices should be designed to be as small and light-weight as possible. A rule of thumb is to make the device with a specific gravity of 1 so that it seems like an extension of your own body. Thus, mobile devices tend to go for smaller, lower mass storage products such as smaller form factor hard disk drives or flash memory. Today hard disk drives in the 2.5-inch form factor are used for mobile consumer applications. A 2.5-inch disk drives are also used in most laptop computers.

A 3.5-inch and 2.5-inch drives are being used in mobile external storage devices that can be connected to laptops and other products. In the last 10 years, 1.8-inch hard disk drives were used for higher-capacity MP3 and video players as well as HDD-based video cameras. 1-inch hard disk drives and smaller were used in some compact MP3 and video players as well as in external storage applications such as HDD-based CompactFlash form factor devices for very high-resolution photographs. The last 1.8-inch HDDs were shipped in 2015, and the last 1-inch HDD were shipped in 2007.

Small form factor hard disk drives used in mobile applications must be durable to avoid issues of shock and vibration that are common in mobile uses. They also must have power saving modes that turn the disk drive partially off when the drive is not being actively read from or written to. If the drive were left on with the disk spinning, it would rapidly use up the power stored in current battery power sources.

Hard disk drive manufacturers have responded to the need for shock resistance by parking the head of the disk when the disk isn't spinning. They also create design rules leaving sufficient sway space around the drive to avoid shock, using material that can absorb vibration as well as prevent vibration due to the mechanical motion of the disk drive and by using sensors that can detect motion of the hard disk drive that mean the drive is falling. In that case the drive can be powered down, and the heads retracted off the disk surface before the host device and the small hard disk drive reaches the ground.

At the time of this writing, thicker 2.5-inch drives with additional disks having 5 TB capacity were available on the market (although thin HDDs that fit into laptops had up to 2 TB capacity).

Hard disk drives are also increasingly being used in automotive applications for navigation systems as well as entertainment. In this environment, both vibration and the temperature range for operating and nonoperating conditions are extreme. Disk drives designed for the automobile market are reduced in storage capacity from equivalent drives used in other applications. This is because by reducing the track density (and to a lesser extent the linear density of the storage in the hard disk drive), the effects of track mis-registration and runout due to the changes in temperature and vibration are reduced. Thus, digital storage capacity is traded off for a more robust hard disk drive.

Static consumer applications include digital video recorders (DVRs, also known as personal video recorders or PVRs), desktop computers used at home including home media center computers, and external storage boxes used for backup, or if network capable, for sharing files. *Static consumer applications* are those where the product containing the storage device is not moved often; hence it is "static." Most static consumer applications use 3.5-inch hard disk drives since these storage products provide the highest storage capacity for the lowest price.

At the time of this writing, 3.5-inch hard disk drives were available with 14 TB of storage capacity on 5 disks (although these high capacity drives are targeted for data center applications). 20 TB and even 40 TB 3.5-inch drives are projected within the next few years.

Much of the content in a home ends up ultimately on static storage devices. Even most mobile consumer products use a copy of MP3 or video file that was copied from a master version that is kept on a more static storage product, either in the cloud (a remote data center), in a desktop, or in a laptop computer.

Because backup is becoming increasingly important in protecting irreplaceable personal content such as digital photographs or personal videos, there are also often multiple copies of content within a home (for instance, the master copy, a copy on a mobile device, and a backup copy on an external storage device in case the drive with the master copy fails). A *home disaster recovery* market for backing up home content through the cloud has become popular with many consumers. Thus, if the house burns down and all copies of a family's digital photographs and personal videos in the home are destroyed, the remote copies are still available and thus not lost.

Table 2.2 Typical specifications for hard disk drives used in common consumer product

Application	Type	Capacity (GB)	Power (peak, W)	Acoustical noise (dB)	Max. shock (operating, G)	Temp. range (C)
Car	65 mm HDD	200	6.5	24	800	−20 to +85
Home	65 mm HDD	500	5.5	26	300	+5 to +55
	95 mm HDD	2000	6.2	29	55	+5 to +55

With the storage capacity of smaller form factor HDDs increasing with time (by at least 15% annually), 2.5-inch HDDs may displace 3.5-inch hard disk drives for some applications, particularly where the size, weight, heat generation, and power consumption are important. A smaller form factor disk drive is better at all of these factors vs. a 3.5-inch disk drive at the expense of lower storage capacity. Thus DVRs with 2.5-inch hard disk drives can be used for lower resolution content and even desktop computer with 2.5-inch disk drives that will actually fit on your desk are now available.

3.5-inch disk drives will continue to be used for higher storage capacity products and also for external storage devices for backup as well as mass digital storage capacity (for instance, to expand the storage available on a DVR so that older shows don't need to be erased to make room for new shows). Table 2.2 shows typical specifications for the common hard disk drive form factors used in consumer products.

2.7 The Cost of Manufacturing a Hard Disk Drive

Hard disk drives have some standard components that are used no matter what the application or form factor. They all have a *spindle motor* to rotate the disk, they all have a *head* attached to a *rotary actuator* to place the head on the data that is read or written from the disk drive, and they all have at least one head and one disk. All have a mechanical housing to protect the hard disk drive components from the outside environment, and they all have electronics on a circuit board that takes the analog voltages from or to the head to read or write information onto the magnetic disks, and controls the motors and the interfaces conveying data back and forth from the hard disk drive to the host device. In addition, hard disk drives have *firmware* that is loaded to give the drive its commands and basic functionality, and they all go through a manufacturing, burn-in, and test process. Firmware is software that is embedded in a device like a hard disk drive.

All of these steps add cost to a hard disk drive. For a 3.5-inch hard disk drive, the basic cost of building a minimum capacity hard disk drive (one-disk and one-head) is in the range of $30. To this cost of materials and direct manufacturing, labor must be added as well as overhead for the facilities and other corporate functions. A break-even price of about $35 for a minimum capacity hard disk drive is likely.

To make money, a drive company must sell this drive for some markup (say 20%), and then the *original equipment manufacturer* (OEM), integrator or the drive distributor or retailer can easily add another 20% or more on top of that. Thus the overall minimum price to the consumer of a minimum hard disk drive in the application is about $50.

2.8 Disk Drive External Interfaces

The external interface to a hard disk drive connects the hard disk drive to the host computer. The interface passes commands and data back and forth between the disk drive and the host. Associated with the physical connection or communication layer, there are higher-level functions that are associated with the interface called higher-level communication layers. For sophisticated communication systems such as Ethernet, there are several higher-level layers that can provide very sophisticated service functions. In the case of today's hard disk drives used in most consumer devices, the electrical interface does not support all the high-level communication layers that true networking architectures such as Ethernet do, but they do support many types of commands and some simple local storage shared disk drive connections. Note that major HDD manufacturers have produced some HDDs with native Ethernet connectivity, but these are focused at present, on enterprise applications.

We will look at the history and development of hard disk drive interfaces. The three major categories of hard disk drive interface in use today are the (AT Attached—as in PC AT in the early 1980s) *ATA*-interface, the *Small Computer System Interface* (SCSI), and the *Fibre Channel* interface. Disk drives and disk drive-based storage systems can be used as *Direct Attached Storage* (DAS) where the storage device directly interfaces with the host or through a network where communication between the storage device and the client is through an intermediary network (which usually has several communication levels) and associated switching or routing hardware.

SCSI and Fibre Channel interfaces are used on high performance hard disk drives used in higher-end direct connect and networked storage arrays (collections of several disk drives in a box that together act as a sophisticated storage system usually using *RAID* or distributed file system architectures) to create larger storage capacity systems with higher I/O performance. Higher I/O performance enables faster data rates between the storage array and the host, while RAID provides the ability to deal with one, two, or sometimes up to three failed disk drives at one time without losing data.

SCSI and Fibre Channel disk drives have higher data rates than ATA-interface disk drives due to generally higher RPM (currently up to 15,000 RPM) and higher performance rotary positioner components used to place the recording and playback head above the disk surface. In order to achieve these high RPMs, these drives use several small form factor disks. The disk size limits the overall storage capacity of SCSI and Fibre Channel disk drives.

Most SCSI and Fibre Channel disk drives are sold for use in servers, workstations, or in storage arrays. Because SCSI and Fibre Channel hard disk drives do not have direct consumer applications today, we will spend less time on them than with ATA-interface hard disk drives.

Parallel interfaces use many parallel signal paths where the data is sent down all the signal paths at one time and put together at the ends of the paths. Because of small differences in effective path length between the various paths in a parallel connection, the data flowing in parallel down the signal paths gets out of phase. In order to reconstruct the data at the ends of the interface, these phase differences must be taken into account. Correcting the "skew" between the various signal paths in a parallel connector becomes harder as the data rate increases. This limits the data rate that parallel interfaces can carry.

Serial interfaces do not have the "skew" issue of parallel path connectors, and with the advent of fast electronics for serializing and deserializing data, it is possible to carry very high data rates in a serial interface.

The parallel ATA-interface developed with the 1984 PC AT but was formalized as a in 1986. This specification was the joint result of work by Control Data Corporation (Imprimis), Western Digital and Compaq Computer. This interface has been called *IDE* by some companies and ATA by others. The first standard document on the ATA-interface came out in 1989. Until 2005 all disk drives used various generations of the parallel ATA-interface. Because of the "skew" issues discussed earlier, the parallel ATA-interface limited the data rate of ATA.

In the early 2000s, the ATA ANSI committee began serious work to come out with a serial interface standard capable of higher data rates. The initial serial ATA standard (SATA I) released in 2004 allowed up to 1.5 Gigabit (or billion bits) per second (Gbps) or 150 megabytes per second (MBps) data rates. The 3.3 specification now allows up to 16.0 Gbps (1600 MBps data rates, making the SATA interface one of the fastest available), particularly to small system developers. SATA 3.3 also allows new SSD form factors (like M.2) as well as shingled magnetic recording support.

The SAS (Serial SCSI) interface was designed to be compatible with SATA for a properly designed common communication backplane. Thus, higher performance and higher priced but lower capacity SAS drives could be combined with higher-capacity, lower priced, and lower performance SATA drives in a common storage system. SAS-3 supports up to 12 Gbps (or 1200 MBps) data rates. SAS-4 will support 22.5 Gbps data rates.

Figure 2.9 compares images of the older parallel "ribbon" connectors used for parallel ATA connections and the current SATA connectors. The serial interface connector is much easier to use than the parallel connector and also more reliable since it was easy to plug the parallel connectors in slightly askew so all the parallel connections were not made. The serial connector is also lighter and more flexible, making it easier to work with.

SATA interface disk drives are the most numerous disk drives produced in the world. They are used in desktop, laptop, and most consumer electronics applications. SATA interface disk drives used in low-cost storage arrays are also the fastest growing portion of the enterprise storage market. SATA disk drives enjoy advantages due to mass production and thus lowering of the cost of components and wide availability of suppliers.

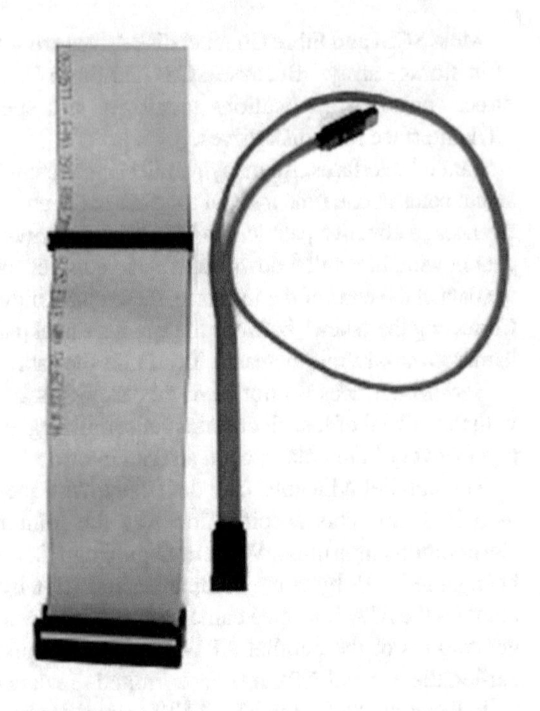

Fig. 2.9 Comparison of parallel ATA connector with a SATA connector

In addition to simplification of the interface with serial connectors, SATA also includes more sophisticated commands than the older ATA storage standards. There are many new commands in the SATA specification that are focused on multiple device connections, external or e-SATA and particular functions associated with common consumer functions such as support for multiple streams of content (such as for parallel write and read on a digital video recorder). SATA disk drives offer a low-cost high-performance interface for consumer applications that need the storage that a hard disk drive can provide.

2.9 Hard Disk Drive Technology Development

The storage capacity of a hard disk drive is the product of the total disk surface area available for data on the drive times the average *areal density* of the information storage on the disks minus any loss of user storage capacity due to formatting of the hard disk drive and other nonuser accessible data. The growth in the storage capacity of hard disk drives has been enabled primarily by the growth in the storage areal density. The areal density of information storage on a disk drive is the product of the average linear density that information can be recorded on a disk track and the average density of the data tracks. Increases in areal density are dependent upon scientific breakthroughs and engineering integration of technologies that allow higher linear densities and track densities on the hard disk drives.

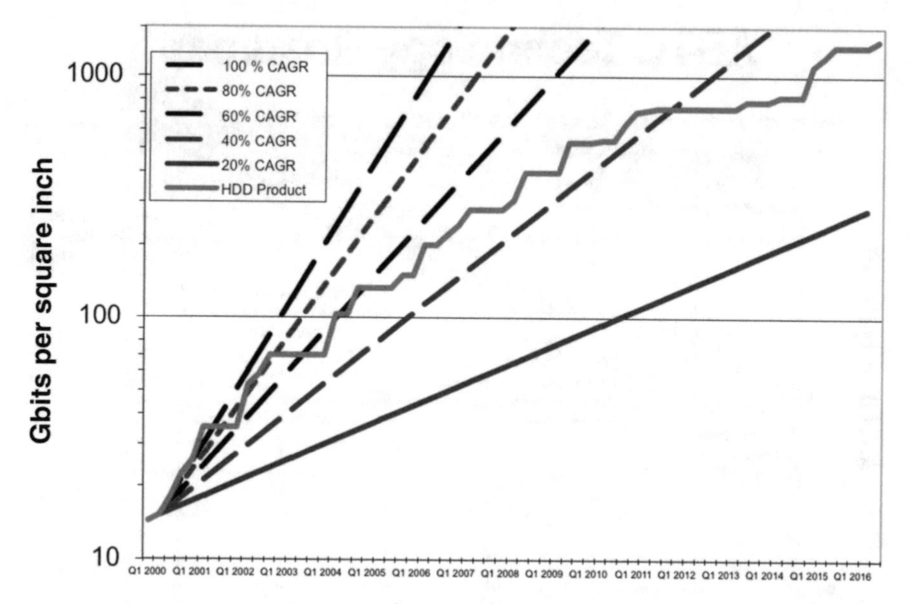

Fig. 2.10 Comparison of product announcement trends vs. areal density annual growth rates

Through the years, progress in HDD technology has been measured through several parameters such as data rates, HDD reliability, shock resistance, and other characteristics. The most significant characteristic that has driven the growth of HDD use in so many new applications, however, is the amount of data that can be packed into a hard disk drive with a finite number of heads and disks. By increasing the storage capacity of a hard disk drive using the same number of components, it has been possible to create a cost per GB of storage that few other storage technologies can match. Some storage technologies such as magnetic tape and some optical recording have a low-cost per GB too, but they do not have the combination of attractive performance and convenience characteristics that have favored HDDs.

At the time of writing, areal densities of shipping disk products exceeded 200.2 Gb/cm^2 [1300 Gb/in^2] with areal density demonstrations exceeding 248.2 Gb/cm^2 [1600 Gb/in^2]. Recent technical advances suggest storage areal densities over the next decade will attain or surpass 1055 Gb/cm^2 [10,000 Gb/in^2]. Other critical enablers for mass storage's continued advances in capacity and performance are perpendicular recording, *heat-assisted magnetic recording* (HAMR), *microwave-assisted magnetic recording* (MAMR), *micro-electromechanical systems* (MEMS) technology, and *patterned media*. Perpendicular recording HDDs started shipping in 2006 and is used in all HDDs today.

Figure 2.10 shows a comparison of the actual quarter-by-quarter logarithmic areal density product announcement curve vs. several *cumulative annual capacity growth rate* (CAGR) curves starting from calendar first quarter 2000 (Q1 2000). The CAGR curves show the expected annual increase in areal density if the same rate of increase happens annually. As can be seen from these figures, it is possible to

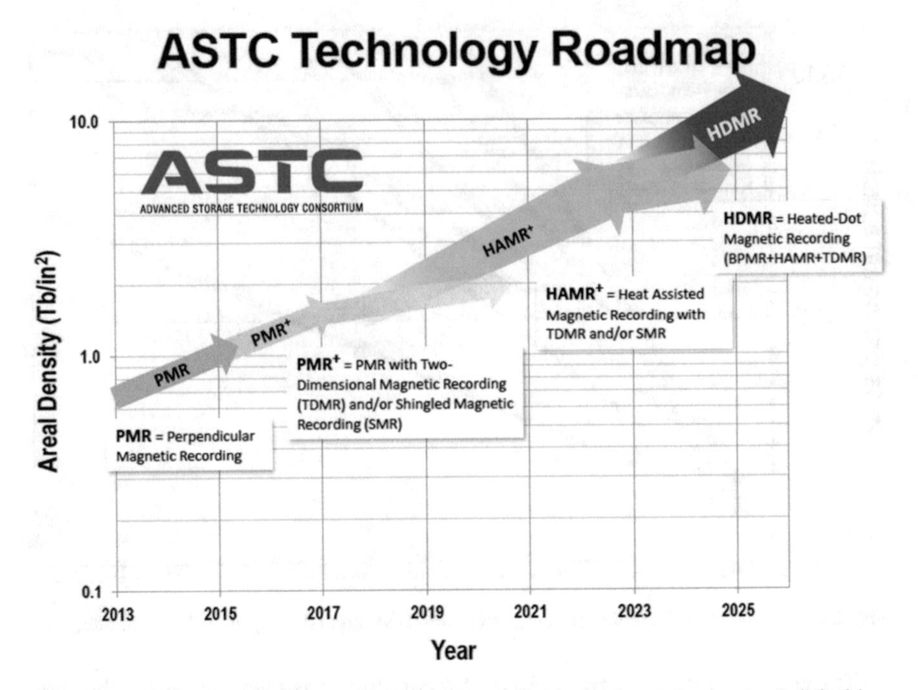

Fig. 2.11 Projected areal densities and recording technologies in production hard disk drives (From the ASTC in 2016)

get a partial fit of the actual announcement data to one or another CAGR curves, but the fit is never that good and doesn't represent the rich complexity of actual technology development. The following are some observations on the introductions of new storage technologies based upon this example of hard disk drives:

- Technology introductions occur erratically in a monotonically increasing fashion, driven by the pace of technical discovery as well as introductions timed for maximum advantage to the introducing company.
- Once a technology is introduced, it must go through a learning cycle until yield and performance issues are resolved and then follow a rapid adoption that displaces other technologies.
- There may be more than one approach that creates at least a short-term solution to a given technological problem.

Figure 2.11 shows broad expectations of the development of magnetic recording technology over the next few years based upon projections from the Advanced Storage Technology Consortium, which includes all the hard disk drive manufacturers (Seagate Technology, Western Digital, and Toshiba).

With the projected growth of product areal density, by 2025, we could see hard drives with the following capacities for the two current HDD form factors:

- 50+ TB 2.5-inch drive
- 100+ TB 3.5-inch drive

Hard disk drive prices declined rapidly in price through the 1990's, forcing large-scale consolidation in the industry but also opening many markets to the use of hard disk drives, such as consumer electronics. However, since 2004 the average price of hard disk drives has declined much more slowly. Still, with the average storage capacity increasing and with a 5% or so average drive price decline, we could see 3.5-inch disk drives with prices per TB of $4.00 by 2025. No other storage technologies except magnetic tape can claim such economical digital storage.

At the same time that areal density has increased so dramatically (say 6000 times over the last 20 years), the time it takes to get to data on the disk drive has increased. The time to data on a hard disk drive is a function of the linear density of data on the hard disk drive tracks, the rotational speed (*RPM*) of the disks, and the layout of data on the hard disk drive.

Hard disk drives are increasing in reliability, driven by many factors. Hard disk drives are using more intelligent electronics and control commands, enabled by more sophisticated electronics on these devices. The inclusion of shock sensors and accelerometers are also enabling disk drives to detect when they are dropped so if the drive is in operation, the heads can be taken off of the disks to prevent head slap and loss of data. All of these factors as well as better understanding of how to build a hard disk drive into a CE device are making hard disk drives more reliable.

Users of storage would like to get to their data faster as well as have lots of it stored economically, and this will lead to some changes in hard disk drives. A hybrid hard disk drive is a hard disk drive with a flash cache inside the drive that holds information that is to be written to the hard disk drive or accessed frequently. Some of the operating system boot information can be stored on the flash memory and used to boot the computer faster than would be the case with the hard disk drive alone. Such hybrid hard disk drives have applications in consumer electronics as well, where the flash can serve as write or read cache to conserve power.

This brings up an important point in the use of digital storage in consumer devices. As the amount of digital data on consumer products increases and as this data is written and erased, the data on the storage device can become fragmented across the disk. This can increase the time to access the data over time and may make power-saving schemes ineffective. For this reason, consumer products must include some internal storage intelligence and the capability to run management functions such as disk defragmentation in the background to preserve the performance and power usage of the consumer device over time. In general, as the electronics involved in building a hard disk drive become more sophisticated and open more real estate on the hard disk drive circuit board, additional intelligence can be built into these storage devices. Some important steps in this development are SMART capabilities and CE-commands in the ATA specification and hybrid drives.

If the trend to greater intelligence in hard disk drives continues, we may see hard disk drives that contain some file system capability, allowing them to know their own contents. Capabilities such as on-board drive searches could be done with such intelligent disk drives. If there are several intelligent disk drives, then such searches could be done in tandem making a very fast search process possible.

Internal file systems or even *object storage* on the hard drives could also allow more intelligent management of the disk drive and the data that it contains. Object storage differs from file systems and block storage. Objects include the data, a variable amount of metadata, and a globally unique identifier. Object storage allows the retention of large amounts of unstructured data. This type of storage system has become common in large cloud data centers.

More sophisticated communication capability could also be built into hard disk drives giving them new wired and wireless communication capability. Many consumer functions could even be designed into hard disk drives, making hard disk drive manufacturers *Original Design Manufacturers* (ODM) for consumer electronics companies. This could save money by eliminating a CE product circuit board and a separate integration of CE and storage electronics and commands. If done with open standards, this could result in faster CE product designs and time to market and also better overall performance and CE product reliability. The development of thinking storage devices that are intelligent nodes on a home and mobile network is not far off.

2.10 Chapter Summary

- Hard disk drives have been a key enabling digital storage technology for over 60 years. Hard disk drives make modern digital computers possible. Today hundreds of millions of these useful devices are manufactured per year for applications ranging from enterprise computing, desktop, and laptop computers to static and mobile consumer applications.
- Hard disk drives have data organized in sectors with the magnetic heads moving across the disk to read and write information while the disk rotates beneath it. The disks are coated with magnetic material, and the heads write and read information on the disks with magnetic fields.
- Disk drives are designed to last several years when used according to their specifications. Different disk drives are designed with different specifications to match the needs of their intended application. Static applications such as digital video recorders use larger 3.5-inch disk drives while mobile applications use smaller form factor disk drives such as 2.5-inch disk drives.
- Disk drives have developed electrical interfaces that provide very fast data transfer speeds. For SATA drives used in many consumer applications, data can be transferred at a native 16 Gbps data rates (SATA 3.3).
- The orientation of the magnetic material in the disk media has changed from in the plane of the disk (longitudinal) to out of the plane of the disk (perpendicular) in order to achieve higher magnetic recording densities.
- Future magnetic recording developments are expected to maintain about a 15% annual increase in the areal density of magnetic recording. Magnetic recording areal density may be able to increase until it is over 10 terabits per square inch. The highest shipping areal density at the time of writing is over 1.3 terabits per square inch. Over 10 times current capacities of the various form factors may be possible using magnetic recording technology.

Chapter 3
Fundamentals of Optical Storage

3.1 Objectives in this Chapter

- Go through the optical storage products now available on the market.
- Understand the basic operation of optical drives and discs.
- Teach the differences between CD, DVD, and Blu-ray disc (BD) optical storage systems, including how each new generation has been able to store more digital content.
- Provide a quick overview of how data is organized on an optical disc.
- Review a history of optical disc form factors.
- Guide the user through the issues and expectations on optical disc reliability and longevity.
- Learn about holographic recording technology and other advanced optical disc technologies.
- Gain insight on what could drive the use of very large capacity optical disc technology.

3.2 Optical Disc Technologies

Although optical discs have served a variety of uses in consumer and computer applications since the 1970s, their popularity has largely been superseded by Internet access to digital content, except for the highest resolution content. However, optical discs are still widely used, due largely to the immense popularity that resulted from their having been based on open standards. Their uses have included physical content distribution (CDs for digital audio and DVDs and Blu-ray discs for video), in

---

The section on optical disc technology owes much to Dr. Richard Zech, Technical Consultant and Expert Witness, specializing in optical data storage technology and to an extensive review by Brian Berg of Berg Software Design.

T.M. Coughlin, *Digital Storage in Consumer Electronics*,
https://doi.org/10.1007/978-3-319-69907-3_3

computers, for backup and archiving of content, as well as file sharing, and in home entertainment systems, to record programs for later viewing. Optical discs are still the most popular physical content distribution media.

Analog Optical Discs

The first commercial optical storage product that met with some success was the laser disc (LD) player,[1] whose media stored 30 or 60 min of analog video and audio on each side of its 30 cm diameter disc. Introduced in 1978, it proved popular for movies for home entertainment systems. A 780 nm wavelength laser housed in an optical pickup unit (OPU) was used for reading the media. Developed in France, the competing and noncompatible Thomson-CSF videodisc system became available in 1980 and was marketed to educational and industrial markets. The videodisc industry is an example of incompatible systems competing for market acceptance. Although videodiscs produced a higher-quality user experience compared to recordable videocassettes, the videodisc market was limited due to the confusion created by the coexistence of both incompatible optical product offerings and incompatible stylus-based product offerings.

Digital Optical Discs: Compact Disc

The digital optical storage industry is an example of the huge success that can come from products based on well-defined standards that are available to the industry. This success was dramatic enough that it is often cited as an example of how a new technology can succeed in the marketplace by way of Philips and Sony having agreed on a compact disc (CD) standard in 1980 and their licensing this technology to competitors so as to avoid repeating the VHS/Betamax standards battle. Known as the "Red Book" because of the color of the cover of the document that defined this read-only (RO) standard, it set forth the physical characteristics and capabilities of the still popular CD audio disc. The Red Book is also the foundation of over a dozen variants of the CD standard, most notably CD-ROM (defined in the "Yellow Book") and also write-once (WO) and rewritable media (defined in versions of the "Orange Book"). A file system for storing CD-ROM data in a manner similar to that used by magnetic disc drives was standardized as ISO 9660 and was later expanded as the Universal Disc Format (UDF) which was also able to accommodate writable CD media.

Digital Optical Discs: DVD

As technology progressed, demand grew for a new standard that was backward-compatible with the CD but which would consolidate CD's many expanded audio and video features with new features, allow for a much higher capacity, and be able to replace videocassettes with higher-quality digital motion picture releases. This led to work on a follow-up optical media format that would better accommodate high-quality video and audio but which ended up being two competing formats from major industry players. A compromise was finally reached for a single format known as DVD which was announced on December 12, 1995. Although sometimes known as "digital video disc" and "digital versatile disc," the name DVD is generally not considered to be an acronym. DVD did not achieve wide commercial success until 1999.

[1] Originally known as MCA DiscoVision in North America, and called LaserVision by Philips.

The DVD Forum published its first specification books for read-only media in 1996: Book A for computer data on DVD-ROM and Book B for movies on DVD-Video. An expanded version of UDF is used on DVD-Video media. Variants on Book D for DVD-R (write-once recordable) were published in 1997 and 2000, Book E for DVD-RAM was published in 1997 and 1999, Book F for DVD-RW (rerecordable) was published in 1999, and DVD-R DL (dual layer) was published in 2005. These recordable formats came to be known as the "dash format." Formed in 1997 by companies including Dell, HP, Philips, and Sony, the DVD + RW Alliance created specifications for the "plus format": DVD + R (recordable), DVD + RW (rewritable), and DVD + R DL (recordable double layer). Only one of the two writable DVD standards (dash or plus) were typically supported by any particular DVD drive until 2004 when drives supporting both formats became available. DVD drives are backward-compatible with most earlier CD formats.

Digital Optical Discs: Blu-Ray Disc

In the years following the introduction and success of the DVD, work started on a follow-up optical media format that would accommodate even higher-quality video and audio and a higher-capacity computer storage media. In "format wars" that had some similarity to what led to the DVD, two formats based on shorter wavelength blue lasers were in competition (Blu-ray and HD-DVD). Based on efforts within the DVD Forum, one format was called HD DVD. The Blu-ray Disc Founders group (founded in 2002) began licensing the Blu-ray disc (BD) format in February 2003, and the first consumer recording device was released later that year. In January 2004, Dell and HP put their support behind BD. In May 2004, the ten members of the Blu-ray Disc Founders group, plus Dell and HP, announced the formation of the Blu-ray Disc Association (BDA). Toshiba and other companies supported the competing HD DVD format.

In June 2004, the DVD Forum approved Version 1.0 of the HD DVD-ROM physical specifications. In August 2004, the BDA approved Version 1.0 of the BD-ROM physical specifications. However, it was not until February 2008 that this round of "format wars" ended when Toshiba said it would phase out support for HD DVD. As a result, the industry went with BD as the single next-generation format after DVD. Like the CD and DVD, BD media is a 12 cm (or 8 cm) diameter plastic disc that is 1.2 mm thick. Unlike older optical disc formats, BD can store 1080p high-definition and 2160p ultrahigh-definition video. BD drives are backward-compatible with most earlier CD and DVD data formats.

Optical storage over the past 40 years has provided a multiplicity of storage solutions. The range spans the first analog video disc and 12″ *write-once* (WO) systems in the 1970s to today's blue laser and DVD drives and media. Optical storage is designed for specific consumer applications (primarily, digital audio and video for *read-only* (RO) and *rewritable* CD, DVD, and Blu-ray media). Strict media standards (specifications) permit specific applications to be implemented by means of signal processing, logical and applications level software, and packaging.

Other Forms of Optical Discs

While CD, DVD, and BD media and drives are familiar to the consumer, there are other forms of optical discs and drives that were used from around 1984 until the late 1990s. Media was typically housed in a protective cartridge, and diameters included 3.5″, 5.25″, 12″, and 14″. Many of these used write-once technology (usually called WORM, for write-once, read-many), but they also included various forms of rewritable media as well. This technology is rarely used anymore, but it played an important role in large-scale data storage for banks, government, and large companies. Interesting applications included document scanning by the Library of Congress and images received from the Hubble Space Telescope.

Table 3.1 summarizes the many types of optical disc drive types that have been available over the years, what their media capabilities were, and how their media capabilities grew over time.

Table 3.2 summarizes the many types of optical disc media.

3.3 Basic Operation of an Optical Disc Drive

Figure 3.1 illustrates the schematic operation of playback from an optical recording media.[2] This figure shows the essential elements of an optical drive. Optical discs have land areas and grooves imprinted on the surface that provide a signal to the playback head when the head goes off track. This signal is used by the drive servo system to bring the head back on track.

Another servo system maintains focus of the laser beam in the groove (or on the land). Pits imprinted into the plastic medium during the stamping phase of the replication process cause variations in the laser light reflected off of the disc surface. As the disc rotates under the optical head, signals are decoded into the data that was stored on the disc. Variations in light and dark areas on the optical disc surface are detected and used to position the head on the disc (focus and tracking), as well as to read the data off of the surface.

Figure 3.2 shows a circuit board from an optical drive.[3] The typical circuit board configuration is a multilayer board with tightly packed *surface-mount technology* (SMT) devices and a System-on-Chip IC to run firmware in order to control all of the electromechanical functions. These products use relatively inexpensive components that are manufactured in high volume.

Figure 3.3 shows some of the main differences between CD and DVD technology as well as the higher-definition BD.[4] The laser beam minimum spot size regulates storage capacity similar to transistor densities in *Moore's law* for

[2] http://content.answers.com/main/content/img/McGrawHill/Encyclopedia/images/CE473300FG0010.gif.

[3] http://www.ixbt.com/optical/nec-hr1100/nec-hr1100-pcb.jpg.

[4] http://www.blu-raydisc.com/en/aboutblu-ray/whatisblu-raydisc/bdvs.dvd.aspx.

Table 3.1 Media capabilities of various optical disc drive (ODD) products

#	ODD drive type	CD media capabilities	DVD media capabilities	BD media capabilities
1	CD-ROM	CD-ROM	–	–
2	DVD-ROM[a]	CD-Rom	DVD-ROM	–
3	Cd-RW	CD-ROM, CD-R, CD-RW	–	–
4	Combo[b]	CD-ROM, CD-R, CD-RW	DVD-ROM	–
5	DVD + RW	CD-ROM, CD-R, CD-RW	DVD-ROM and DVD + RW;and eventually DVD + R	–
6	DVD-RW	CD-ROM, CD-R, CD-RW	DVD-ROM, DVD-R,DVD-RW	–
7	DVD+/−RW	CD-ROM, CD-R, CD-RW	DVD-ROM, DVD-R,DVD-RW, DVD + R, DVD + RW;and eventually DVD + R DL;and eventually DVD-R DL;and eventually DVD-RAM	–
8	BD-RE	CD-ROM, CD-R, CD-RW	DVD-ROM, DVD+/−R,DVD+/−RW, DVD+/−R DL;and eventually DVD-RAM	BD-ROM,BD-R,BD-RE
9	BD-Combo[c]	CD-ROM, CD-R, CD-RW	DVD-ROM, DVD+/−R,DVD+/−RW, DVD+/−R DL;and eventually DVD-RAM	BD-ROM

[a]DVD-ROM ODDs also became able to read recordable CD and DVD media
[b]Combo ODDs could read recordable and read-only DVD media and could read and write CD media
[c]BD-combo ODDs could read recordable and read-only BD media and could read and write CD and DVD media

semiconductors. Conventional CD and DVD optical drives use near-infrared (780 nm) and red (650 nm) lasers, respectively. Near-infrared and red lasers have a longer wavelength than the blue-violet laser light (405 nm) used for BD drives.

A shorter wavelength yields a smaller laser spot when imaged by a lens with a larger *numerical aperture* (NA), 0.85 for BD, which in turn permits a smaller size recorded bit and tighter track pitches and, thus, higher-density recording. CDs and DVDs have nominal storage capacities of 650–800 MB and 4.7 GB, respectively, for a single layer; a double-layer DVD has a capacity of 8.5 GB.

Single-layer BD discs have a nominal storage capacity of 25 GB. Two-layer Blu-ray discs provide 50 GB storage capacity. Blue-violet laser technology has been demonstrated for four-layer and eight-layer discs, yielding 100 GB and 200 GB capacities. Blu-ray prerecorded UHD products are available with up to 100 GB capacity.

Table 3.2 Optical disc media types

#	Media type	Name	Definition
1	CD-ROM	CD read-only memory	Read-only CD medium for digital data, including computer data or various multimedia formats
2	CD-R	CD recordable	Write-once CD medium
3	CD-RW	CD rewritable	Rewritable CD medium
4	DVD-ROM	DVD read-only memory	Read-only DVD medium for digital data, including digital video movie data or computer data
5	DVD + R	DVD plus recordable	Write-once wobble groove-based DVD medium
6	DVD + RW	DVD plus rewritable	Rewritable wobble groove-based DVD medium
7	DVD + R DL	DVD plus double layer	DVD + R with two recordable layers
8	DVD-R	DVD dash recordable	Write-once wobble groove-based DVD medium; also known as DVD-R for general
9	DVD-RW	DVD dash rerecordable	Rerecordable wobble groove-based DVD medium
10	DVD-R DL	DVD dash double layer	DVD-R with two recordable layers
11	DVD-RAM	DVD	Rewritable DVD media whose spiral has stamped headers, thereby negating the need for a wobble groove or formatting before being written randomly
12	BD-ROM	BD read-only memory	Read-only BD medium for digital data, including digital video movie data or computer data
13	BD-R	BD recordable	Write-once BD medium written in increments of 65,536 bytes
14	BD-RE	BD rewritable	Rewritable BD medium

3.4 How Data is Organized on an Optical Disc

Conventional optical discs store information by making contrast variation on a disc surface that can be detected as reflected light from a read laser as the disc is rotated beneath it. For read-only discs, such as those used for content distribution, the optical contrast is made by imprinting pits in the disc surface that create light and dark regions at the pit edges. These edges can be detected by reflecting laser light off the surface of the optical disc. The pit edge transitions that represent the digital data stored on the read-only optical disc are organized in a spiral track.

The tracks are broken up into sectors (similar to those on hard disc drives) of data with servo, error correction, and other nonuser data between the data sectors. The size of the sectors is a fixed number of user data bytes such as 2048 bytes. The physical location of data is referred to as the physical sector. The determination of the physical location of data is the job of the electronics in the data storage device. The data location as seen by the host that the storage device is attached to is referred to as the logical sector. Logical sectors "virtualize" the physical sectors and there is no necessary correlation between the physical sectors and the logical sectors. This "virtualization" of data is very common in storage devices.

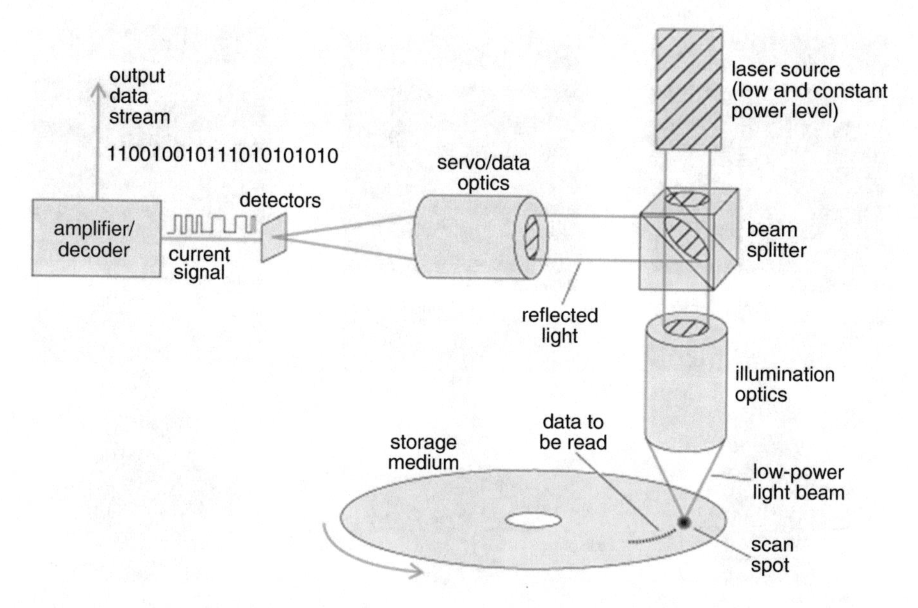

Fig. 3.1 Playback operation of an optical disc

Fig. 3.2 Circuit board of an optical playback drive

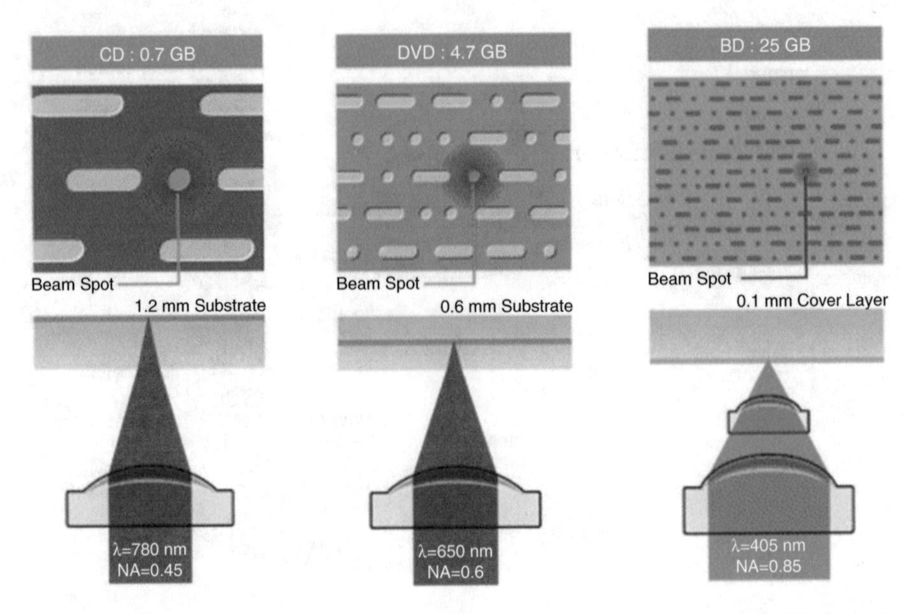

Fig. 3.3 Difference in spot size, pit size, and numerical aperture for various optical drive formats

Data files that are made up of consecutive logical sectors may be broken up in physical sectors that can be located anywhere on the medium. For rewritable media such as hard drive magnetic media and some optical discs, the physical and logical sectors may be quite different due to reallocated sectors (required to replace bad sectors, as defects develop or due to rearrangement of sectors of data as the storage medium fills up). For read-only media, the physical sectors and the logical sectors will be pretty close to the same, since it is usually more efficient to read consecutive physical sectors.

Optical discs vary in storage capacity because of differences in the track and linear densities, similar to the way areal density is increased on hard disc drives. Optical discs differ from hard drives though in that they can record information on layers in the depth of the plastic optical disc. These are accessible by focusing the read laser beam to different focal depths in the media where information is recorded. Thus, for instance, a BD disc can have 25 GB capacity or 50 GB capacity depending upon whether a single layer or double layer is used. Optical read-only disc layers are made by stacking layers of imprinted plastic media together to make a thicker plastic optical disc.

CDs and DVDs use near-infrared and red lasers, respectively, which have longer wavelengths than the blue-violet lasers used in BD discs. The longer the light wavelength, the larger the contrast features that can be created and resolved. DVDs use a somewhat shorter wavelength light than CDs, so they can use smaller pits and also denser tracks.

Likewise BD products have even smaller pits and denser tracks giving them a single-layer capacity of about 25 GB. Two layers are available today with 50 GB capacity. Three or four layers are possible. With multiple layers there are issues associated with signal loss from optical absorption of the laser light by the thickness of the medium between the laser and the detected layer. The effective limits of the layers that can be used are a function of the difficulty in manufacturing multiple layers, spherical aberration compensation, and the optical losses due to absorption when reading the different layers.

3.5 Optical Disc Form Factors

Optical disc drive form factors include half-height as used in desktop computers, as well as slim (12.7 mm), ultra-slim (9.5 mm), and internal slim (7.5 mm) as used in notebook computers.

There have been many other form factors that have been introduced over the years, including the mini-DVDs used in some older camcorders and mini-CDs that were popular in Japan. DataPlay, Sony, and Philips introduced various optical formats, including ones about an inch in diameter. All of these alternative optical disc formats are mostly obsolete.

The vast majority of optical discs are read-only discs used for software or entertainment content distribution. Optical discs are often the cheapest and fastest way to receive fixed content, such as music or video. Some companies use small form factor optical discs (as well as the larger discs) in portable player devices.

When introduced, optical storage technologies are rather expensive, and even in the consumer market, drives initially sell for $1000, or more. This price drops rapidly with production volume and yields. Once the technology ramps up, the price of drives drops by 50% per year for the first few years. Blu-ray players have been sold for about $40 retail in recent years with a small profit margin.

In volume, optical media drops dramatically in price. Production costs of a high-volume medium are a few cents per disc. Total retail production costs, including replication and packaging, are in the range of $1.50–$2.00 each. If such product sells for $15 each, you can see that the profit margin is pretty good for prepackaged content.

3.6 Optical Disc Reliability

Optical storage products are complex. Fortunately, each new generation is able to build on its predecessors. Moreover, CD, DVD, and Blu-ray disc products are assembled from only a few subassemblies. The *optical pickup unit* (OPU) and all mechanical elements (including the spindle and load/unload motors) are preassembled into what is called an *Optical Mechanical Assembly* (OMA). An electronics main board handles the servo, read/write channel, and controller functions. Final assembly consists of attaching the OMA to the chassis, attaching the main board to the OMA, attaching connectors, and statistical burn-in and test.

The OMA used in a CD, DVD, and BD player assigns all control, electronic and electrical functions to a System-on-Chip (SoC), which is a single integrated circuit chip that is able to control all of the functionality of the player. A number of Asian companies (e.g., MediaTek, Sanyo, Sharp, and Sony) sell kits that permit design and assembly companies to both customize and rapidly bring to market CD, DVD, and BD products. This approach has greatly improved quality and reliability over the past 10 years. This is because complex assembly functions, for example, optical head alignment, are completed by professionals in well-equipped assembly areas.

Optical data storage products are amazing for several reasons: (1) a high degree of complexity reduced to ordinary assembly procedures, (2) extremely low prices relative to performance and function, and (3) a surprisingly high level of reliability. Some studies of customer service/repair records indicate that less than 0.1% of optical storage products have a hard failure, if they survive infant mortality.

With reasonable treatment and usage, most optical storage products will give good service for at least 5 years. MTBF for all optical storage products is estimated to be over 100,000 power-on hours. Optical storage reliability should not be confused with return rate, which is in the 3–5%. This is mainly a "buyer's remorse" issue, supported by easy return policies of many retailers (especially in the US). Also, much emphasis has been given to optical media quality and reliability.

Some reports suggest that the archival life of recordable discs is only 3–5 years, and magnetic tape should be used for archiving. This may in fact be true for the optical media bought for $0.05 per disc; cheap discs are only one level removed from rejects.[5] High-quality media produced by reputable companies can be expected to last at least 25 years, if not abused.

3.7 Holographic Recording

Lasers with wavelengths shorter than that of blue-violet lasers (405 nm) are possible, but it is becoming more and more difficult to move to shorter wavelengths using known technology (wavelengths less than about 380 nm fall into the UV regime, and UV light is destructive to most plastics). Thus, other methods to increase optical recording storage density are needed besides decreasing the wavelength of the light used to record and read. Holographic recording uses a very different approach to record and play back data (analog, interferometer). Figure 3.4 shows how information is written on a photosensitive recording medium by mixing a "reference" laser beam and a "data" beam.

The data beam is created by means of modulation with an electro-optical device called a spatial light modulator (SLM), which forms a "page" of digital data. The

[5] Although not often discussed in public forums, major optical media manufacturers such as Ritek and CMC Magnetics have the ability to produce discs that fall into A, B, and C categories. A Grade is highest quality, B Grade is intermediate quality, and C Grade lowest quality (often used as market loss leaders under obscure brand names).

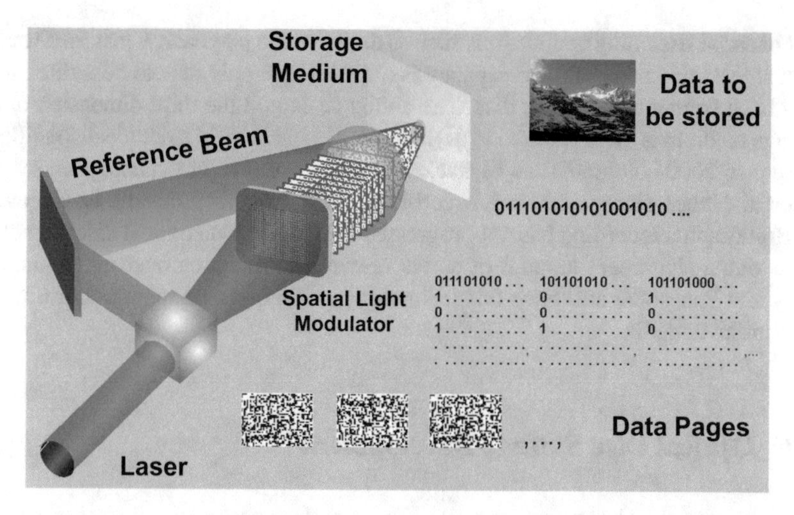

Fig. 3.4 Writing information on a holographic storage medium (Source: InPhase Technology)

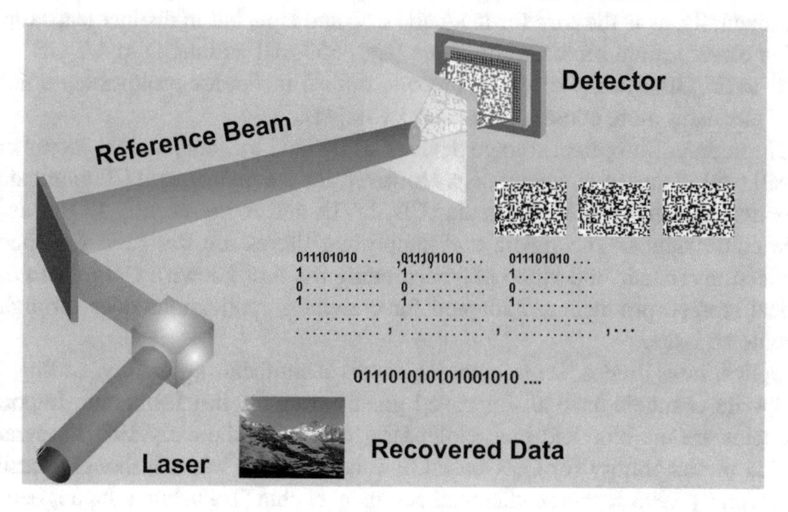

Fig. 3.5 Reading information from a holographic storage medium (Source: InPhase Technology)

mixing of the reference and data beams form interference fringes throughout the volume of the storage medium that is an analog encoding process. The recorded data are reconstructed by illuminating the storage medium with only the reference beam (now called the "reconstruction" beam). Figure 3.5 illustrates hologram reconstruction of a page of data.

Two major advantages of holographic storage are massively parallel write and read and three-dimensional access to the storage medium. Rather than being recorded or read back as a stream of bits, holographically recorded data are read back in "pages" of data that can be several MBs in size. This, in principle, permits write and read data transfer rates to be very high.

For serial streaming applications such as digital video playback, a fast FIFO buffer with at least two pages of data capacity is required to supply data to be written to or read back from a holographic disc. The ability to exploit the third dimension of the storage medium is often referred to as (hologram) "stacking." Under ideal conditions, more than 2000 holograms can be stacked in a fixed volume of the hologram storage material. Hence, storage densities exceeding 1 TB/in^2 can theoretically be achieved.

Holographic recording has not progressed into production optical discs although there was a significant amount of active research in this area over the years, and Sony and Panasonic still have this technology on their roadmap as a future write-once archive media.

3.8 Optical Disc Storage Development

Capacity and throughput for optical storage will continue to improve, though it occurs more slowly than for magnetic storage. Optical disc capacity increases not incrementally, as is the case for magnetic disc and tape, but in distinct leaps (application driven) often separated by years (e.g., 650 MB audio CD to 4.7 GB video DVD to 25 GB Blu-ray discs). Moreover, optical media are removable, which in itself mandates more conservative capacity targets.

Throughput for optical storage devices, as defined by data rate and access time, is well behind that of magnetic disc. However, for the mainstream CE applications of optical disc hardware and media (CD, DVD, and more recently BD), this has little consequence. The media and supporting drives are designed for specific applications (music and video playback being the best known). Only when these optical storage products are adapted for computer applications does throughput become an issue.

Optics, laser diodes, servo controls, media manufacturing quality, coding, and read/write channels have all improved greatly over the last few years. Important examples are the blue and blue-violet laser diodes. In the early 1990s, operation outside the laboratory for GaN-based blue laser diodes was not thought feasible. However, by 1995 Japanese chemical company Nichia (Tokushima, Japan) demonstrated the first stable devices. By 2001, the company was sampling 405 nm, 5 mW blue-violet laser diode kits for $5000. Today, 405 nm 35 mW laser diodes sell for less than $10 OEM. Laser diodes with continuous wave (CW) output power of 300 mW have been demonstrated by Samsung. Moreover, operating life now exceeds 10,000 h in many cases.

Blu-ray discs (BD) is a successor technology to DVD for video playback and recording and PC/workstation data storage applications. They are application-specific products developed for high-definition television (HDTV) playback and recording but also can be used for computer data storage devices, as was the case for CD and DVD. HDTV requires about four times the bandwidth of standard-definition television (SDTV), implying the need for four times the capacity (this actually depends on the compression algorithm used).

Table 3.3 Comparison of DVD and BD discs

	DVD	Blu-ray disc
Capacity	4.7–8.5 GB	25/100 GB
Data transfer rate (Mbps)	11	36
Laser	Red	Blue-violet
Cover layer	0.6 mm	0.1 mm
Multilayer potential	2 max	8 max
Video compression	MPEG2	MPEG2, MPEG4, WM9
Supporting organization	DVD forum	Blu-ray disc association

For BD, some fine-tuning of the MPEG-2 codec, 25 GB capacity (single layer), and 36 Mbps data rate yields 135 min of HDTV, plus extras (this contrasts with 4.7 GB and 11 Mbps for the DVD SDTV standard). BD supports write-once, read-only, and rewritable discs as well as multilayer discs to increase the storage under the optical head. Capacity is limited by optical absorption, spherical aberration, and SNR of the optical signals through the layers. Double- or triple-layer BD discs are required for UHDTV video.

Increasing storage capacity of optical discs tends to come in discrete and significant jumps to support higher-definition content formats. For instance, 650–800 MB CDs are used for single music album distribution while 4.7 GB DVDs are used for SD movie distribution. 20–100 GB Blu-ray discs enable HD and UHD movie distribution. If physical content distribution continues to be the main use of optical discs, then the growth of optical disc capacities will follow the growth in home (and professional) content resolution.

Table 3.3 compares some of the significant differences between DVD and BD discs.

Figure 3.6 shows a projection for the required data rate and storage capacity to contain various sizes of multimedia objects. As can be seen from this chart, blue-violet laser optical discs will have capacity large enough to contain compressed HDTV and UHDTV (4 K) quality movies, but higher-resolution compressed content, like 8 K video and high-resolution 360-degree video, may require 100 s of GBs, maybe beyond the 200 GB maximum discussed with multilayer blue laser products. UHD 4 K content will be commonplace by 2020 and there even be some 8 K content will be available in some markets (like Japan).

The demand of humans for ever more realistic video experiences will require even higher resolution, higher dynamic range, and greater color gamut requirements. Imagine, for instance, the storage requirements for a 360-degree 3D movie displayed in high dynamic range at 8 K and viewable from any angle. Such products could well become common household products within the first 30 years of this century. The result may be demand for optical physical distribution media with storage capacities of 1 TB or higher (or much higher bandwidth Internet connectivity).

One TB optical discs or larger could be possible with mass-produced holographic discs, large number of optical layers on a disc, or for even higher-frequency optical disc products. Even with the rise of content distribution through the Internet, as long as the resolution requirements increase faster than the bandwidth available to most consumers, there will continue to be demand for physical media.

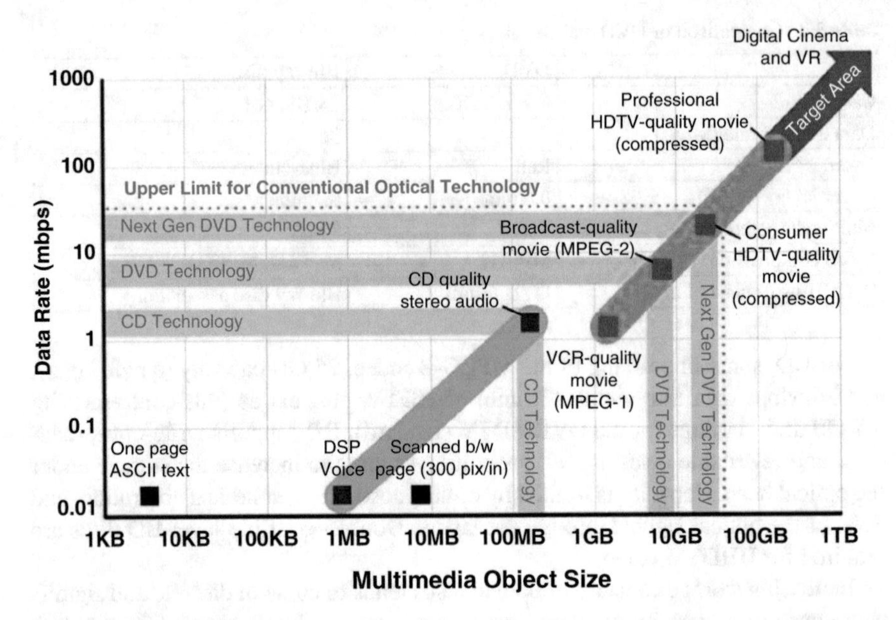

Fig. 3.6 Comparison of storage capacity requirements and data rates for different size multimedia objects (Image courtesy of Telcordia)

The mass production cost of optical media tends to be a few cents per disc. As the storage capacity increases, the cost per unit of storage becomes very low. Moving from DVD to Blu-ray discs dropped the price per GB. The $/GB of optical media will continue to drop if holographic recording or even higher-frequency optical recording moves into consumer products.

3.9 Chapter Summary

- Optical disc storage technology has been around for over 40 years. There are many optical storage products now available on the market. Some of these are professional products for niche markets while others are designed for consumer applications. Many of the consumer optical formats were primarily used for content distribution of music or video.
- Optical disc drives use mirrors and lenses to move laser light around and focus it on the target optical disc for reading or writing. As the wavelength of the laser light decreases, the optical spots that are written or can be read become smaller and so higher-density recording is possible. Thus, the capacity of blue-violet laser BD media is greater than that of near-infrared laser CD and red-laser DVD media.
- Data is written on read-only optical discs as pits on predefined tracks. The optical contrast of the transition from the pit edges provides the signal that contains the digital information. Much like a hard disc drive, the data on the tracks is broken up into sectors with data and servo information in each sector.

- Many small form factors have been used in the 40-plus-year history of optical recording. Smaller form factor discs have been popular in players and in video recorders, especially in Japan.
- Optical discs can have a range of expected lifetimes depending upon the care and materials with which they have been made. High-quality optical discs kept in an ideal environment (temperature and humidity) can be expected to survive for 20 years (probably long after the devices that could read them have disappeared). Optical storage disc drives are generally specified for 100,000 power-on hours.
- Holographic recording can record information in an optical disc format through the volume of the recording media. This technology, although worked on for over 40 years, has never come to production. Although companies like Panasonic and Sony project 1 TB holographic optical discs for archiving, it appears we are years away from consumer products using holographic recording.
- Rich content distribution drives the development and use of new optical recording technologies. Blue-violet laser optical discs have achieved market dominance over older formats.
- UHD 4 K and even 8 K content will require 100 GB and higher-capacity optical disc media. This may lead to very high-layer Blu-ray discs, shorter wavelength lasers, or perhaps technologies such as holographic recording finally finding a mass market. With very high-resolution 360-degree content, a 1 TB disc may well be needed.

Chapter 4
Fundamentals of Flash Memory and Other Solid-State Memory Technologies

4.1 Objectives in this Chapter

- Review the history of flash memory technology.
- Describe the basic flash operations of erase, write, and read.
- Explore the causes and operation of flash memory cell wear.
- Present the differences of NOR and NAND flash memory technologies.
- Understand bit errors in flash memory.
- Find out how to reduce the negative effects of flash cell wear with wear leveling.
- Discover the basics of bad block management on flash memory.
- Review embedded vs. removable flash memory technology.
- Learn how flash file systems make flash data accessible.
- Find out about the advantages and trade-offs of single-level cell memory and multilevel cell memory.
- See how die stacking can be used to increase volumetric density of flash-based devices.
- Explore the various ways that flash reliability can be assured and what sort of useful life can be expected in an application.
- Learn the various flash memory card formats.
- Understand what can be expected in terms of flash memory prices over the next few years.
- Learn about 3D flash and emerging solid-state storage technologies.

4.2 Development and History of Flash Memory[1]

Flash is an extension of the *floating gate* method of manufacturing nonvolatile memory. The first sort of floating gate memory was the *erasable programmable read-only memory* (EPROM), invented in the 1960s but not developed until the 1970s. In EPROM, like in dynamic *random-access memory* (DRAM) and *read-only memory* (ROM), each memory bit was represented by a transistor.

---

[1] Thanks to Jim Handy of Objective Analysis for much of the materials used in this section.

T.M. Coughlin, *Digital Storage in Consumer Electronics*,
https://doi.org/10.1007/978-3-319-69907-3_4

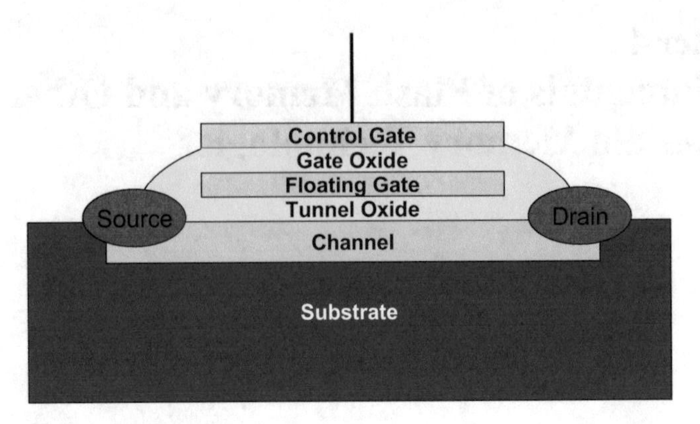

Fig. 4.1 Floating gate flash memory cell

Appendix C gives a list of some of the most significant flash memory manufacturers.

It helps to understand how transistors work if you want to understand these circuits. In a *field-effect transistor*, FET, current flows from the source to the drain. The gate controls how much current flows through the channel (the area between the source and drain). If the gate is unbiased, then the current flows through the channel relatively freely. If a bias is applied to the gate, then the channel depletes—that is, the carriers are moved out of part of the channel—making it seem narrower and limiting the current flow. This principle is key to all FET-based technologies—*metal-oxide semiconductor* (MOS), *complementary metal-oxide semiconductor* (CMOS), *bipolar complementary metal-oxide semiconductor* (BiCMOS), etc. The ability to turn a circuit's current flow on and off allows individual bits to be routed to a data bus.

In a DRAM, each memory bit's transistor uses a capacitor to store the bit. In a ROM, each bit's transistor uses a short or open circuit to represent a bit (either programmed at manufacture by a mask or afterward by a fuse that could be blown). In EPROM, the transistor itself looks like it contains something like a capacitor, but it is actually a second gate that complemented the control gate of each bit's transistor. The bit's transistor actually has two gates, one that is connected to the bit line, and one that was connected to nothing – it "floats". This technology is known as a "floating gate." This concept is used by four technologies: EPROM, *electrically erasable programmable read-only memory* (EEPROM), NOR flash, and NAND.

Figure 4.1 illustrates the cross section of a floating gate. When a bit is programmed, electrons are stored upon the floating gate. This has the effect of offsetting the charge on the control gate of the transistor. If there is no charge upon the floating gate, then the control gate's charge determines whether or not current flows through the channel: A strong charge on the control gate assures that no current flows. A weak charge will allow a strong current to flow through.

The floating gate is capable of storing a charge of its own. This adds to the bias of the control gate. If there is a charge on the floating gate, then no current will flow through the channel whether or not there is a charge on the control gate.

One alternative to the floating gate is the charge trap. These two technologies differ very slightly: A floating gate is usually manufactured using conductive

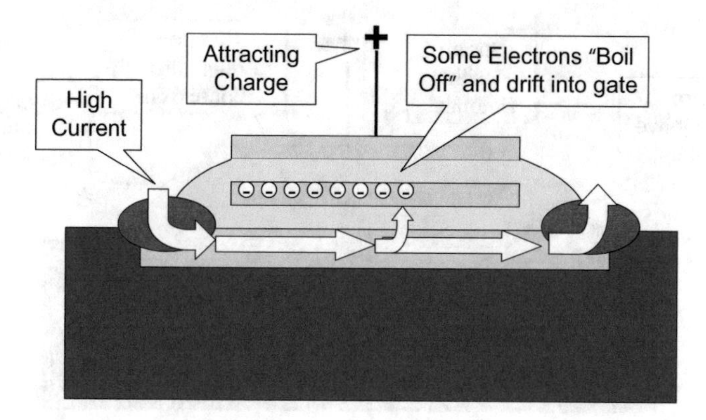

Fig. 4.2 Channel hot electron injection

polysilicon, while a charge trap is manufactured using insulating silicon nitride. Both have advantages and disadvantages. A polysilicon floating gate is easier to move electrons onto and off of, although it takes more energy to do so. A silicon nitride charge trap is easier to manufacture, since the traps from adjacent memory cells don't need to be insulated from each other since they are already formed of an insulator. This simpler form of manufacturing lends itself to the production of many modern 3D NAND flash structures.

4.3 Erasing, Writing, and Reading Flash Memory

The key to using floating gate technology is to put a charge onto the floating gate when it is needed and to take it off when it is no longer needed. There are two methods of getting charges on to and off of the floating gate: *channel hot electron* injection (CHE) and *Fowler-Nordheim* tunneling (FN). Both concepts involve quantum methods, so they will be dealt with here in a very simplistic way.

Channel hot electron injection takes advantage of the fact that electrons are more mobile with high current levels. When CHE is used to add electrons to a floating gate, a high source to drain current is run through the channel. With this high current, a number of electrons are looking for ways to "boil" out of the channel. A positive bias on the control gate attracts electrons from the channel into the floating gate where they become trapped. This adds a bias onto the floating gate (see Fig. 4.2).

Fowler-Nordheim tunneling requires a high voltage (compared to the operating voltage of the chip—usually between 7–12 Volts) to be placed between the source and the control gate of the transistor. If the voltage is sufficient, the electrons "tunnel" through the gate oxide layer and come to rest upon the floating gate. This is illustrated in Fig. 4.3.

In EPROM, the original floating gate technology, the memory was erased through exposure to ultraviolet light. Light rays would allow the electrons trapped in the floating gate to migrate back to the channel. In EEPROM, Fowler-Nordheim

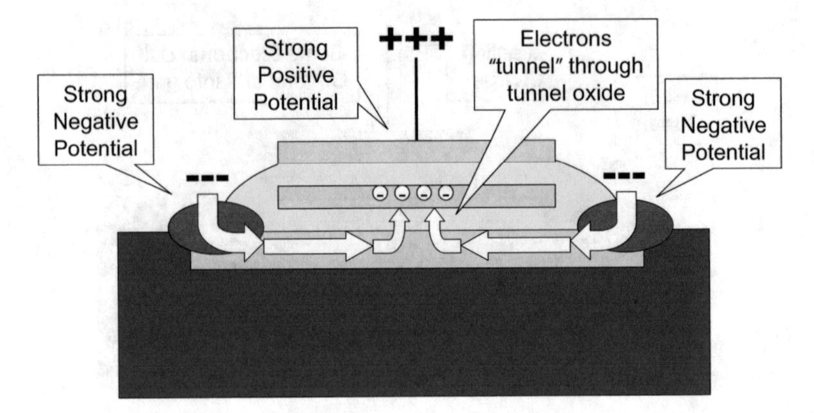

Fig. 4.3 Fowler-Nordheim tunneling

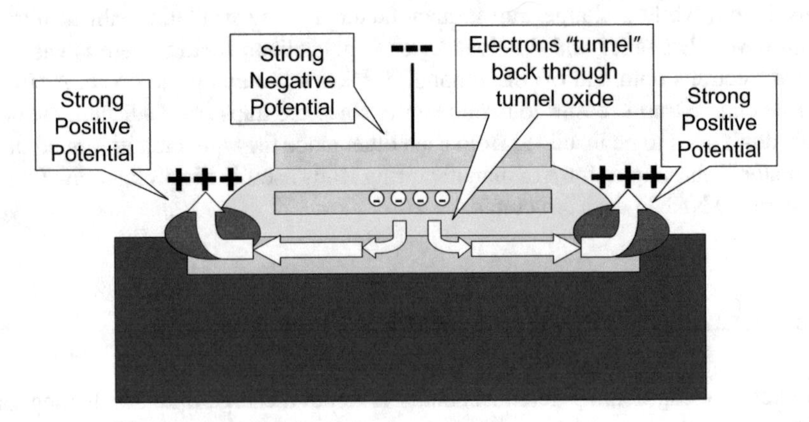

Fig. 4.4 Erasing the bit through tunneling

tunneling was reversed by a second transistor on the memory bit to take electrons back out of the floating gate and put them back into the source where they originally came from. This erasure approach is illustrated in Fig. 4.4. Both NAND and NOR flash technology use the same electronic erasure concept as EEPROM, but rather than add an extra erase transistor to each individual bit cell, a single very large transistor erases all the transistors in a sub-array called a "block." This effectively halves the size of the memory array, providing significant cost savings to designers who don't need to be able to individually erase each bit or byte of the memory.

4.4 Difficulties that Cause "Wear" in Flash Memory

Tunneling electrons migrating through the tunnel oxide sometimes causes difficulties. It is inevitable that electrons from time to time will get trapped in the tunnel oxide. These electrons, once trapped, cannot be removed. This will impact the operation of the cell to a certain degree, depending upon the number of electrons trapped in the oxide: if a lot of electrons are trapped, then there will be a big impact, but a low number of electrons are not likely to cause much impact at all. As the number of trapped electrons in the oxide gets too large, it becomes harder to maintain electrons in the floating gate. Thus the memory cell leaks the stored charge.

The number of electrons trapped in any one cell's tunnel oxide is a function of how the chip is made and more importantly of how many times that particular transistor has been erased and rewritten. A specification has thus been devised to recommend a maximum number of erase/write cycles a memory cell on a certain chip can withstand before a failure is likely to occur, and this specification is called the chip's "endurance."

Typical endurances have been managed up to very high levels over the course of time. For many chips today, endurance is specified as 10^5 erase/write cycles; however, some cost reduction techniques can lower this number to a few hundred cycles. As we will see later, some novel approaches are used to avoid endurance failures that will impact the operation of a system.

4.5 Common Flash Memory Storage Technologies: NOR and NAND

There are two kinds of flash memory, *NOR* and *NAND*. The two terms are names of types of logic gates, the negated "or" function and the negated "and" function. The big difference between the two types of architectures is real estate. NAND has a significantly smaller die size than NOR does. This translates to significant cost savings.

These cost savings come with a trade-off. NAND does not behave like other memories. While NOR, SRAM, and DRAM are random-access devices (the "RAM" part of DRAM and SRAM stands for "random-access memory"), NAND is part random and part serial. Once an address is given to the device, there is a long pause; then that address and several adjacent addresses' data come out in a burst like a machine gun. We will explain why in the next two sections.

Another way that NAND makers squeeze out costs is by taking some shortcuts in assuring data integrity. NAND was designed to replace HDDs back in the 1980s. At that time, some thought that there was a brick wall limiting the capacity of magnetic storage. Once HDDs hit that brick wall, then semiconductor memory could race past it in density and cost. NAND's inventors borrowed a leaf from the HDD designers' book by allowing their design to have occasional errors that would be corrected by external logic. Both HDD designers and NAND designers were able to use this more relaxed approach to data integrity to squeeze more bits onto a given piece of real estate (whether on a magnetic disk or on a silicon chip) than would be possible if absolute integrity were to be maintained.

These two techniques, serial access and lower data integrity, allow NAND die sizes to be less than half the size of their NOR counterparts. Let's look at the two of these in depth.

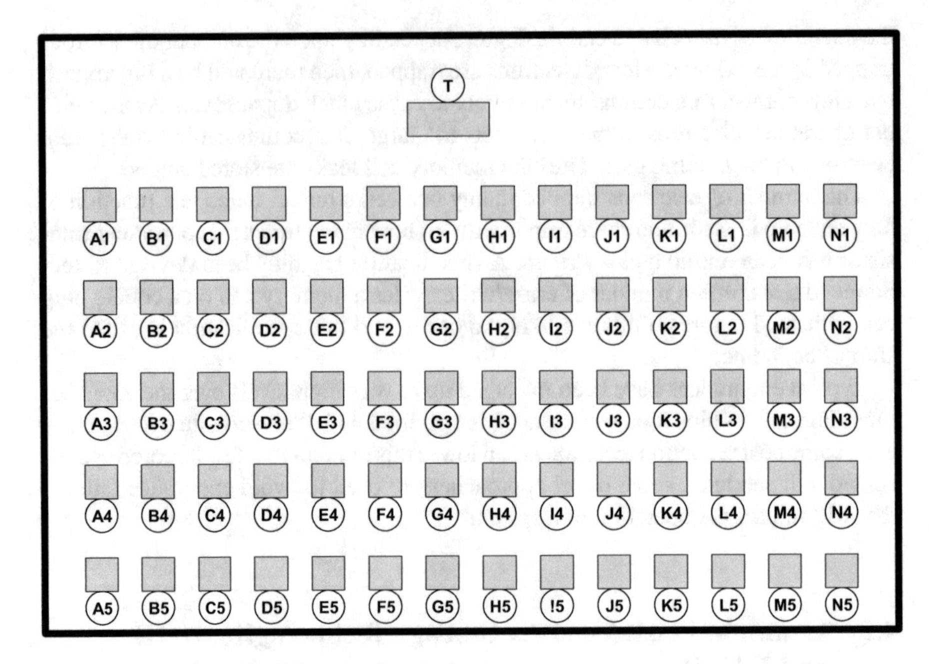

Fig. 4.5 Classroom like NOR. A spacious layout

4.5.1 How Does NOR Memory Work?

We will use a classroom analogy. Picture a class where there is space around all the desks. Each of the desks will represent a bit transistor in the memory, with the students representing data. The teacher represents the system interface that handles the data transaction between the rest of the system and the memory bits.

We have labeled the students in Fig. 4.5 as A1, B1, etc. to indicate the row and column where that student sits. In this example, there are 14 columns (A–N) and five rows (1–5), giving 70 students altogether.

Most memories are designed similarly to this classroom. There are rows and columns and word lines to tell the memory bits which row is being requested and bit lines to access a column within that row.

When a memory bit is requested, it is similar to the teacher calling a student to the front of the room. Say the teacher in this example called for the student B3 to come to the front of the room (Fig. 4.6). That student would simply walk from his desk up the aisle to go to the teacher's desk. All students could get to the teacher's desk in about the same amount of time, and it would not take very long to get there.

This is very similar to the way that a NOR memory works. All data can be accessed rapidly, and all accesses take about the same amount of time. Data can be requested from individual locations in a completely random sequence.

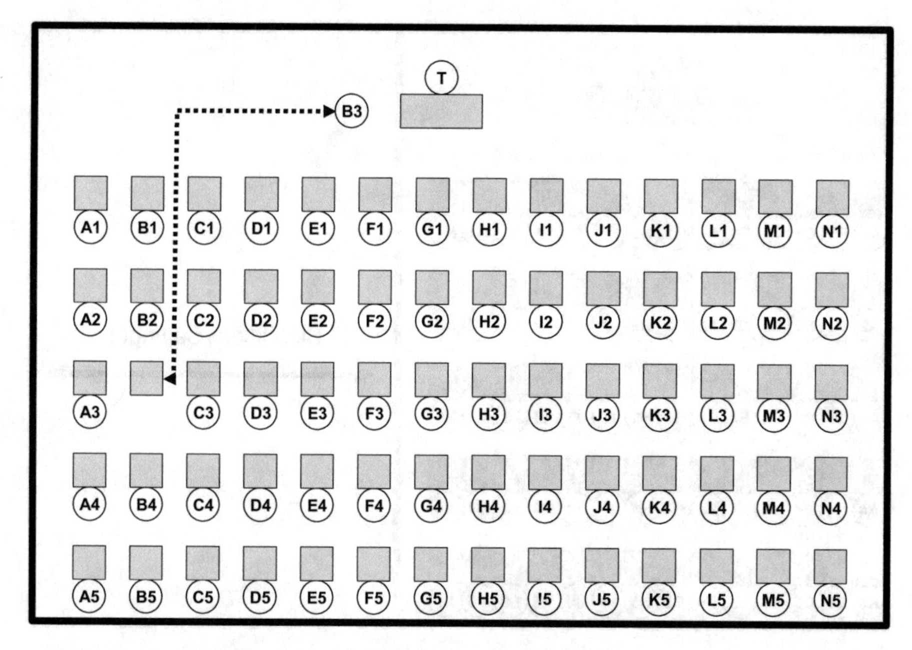

Fig. 4.6 The teacher calls student B3 to the front of the NOR classroom

4.5.2 How Does NAND Memory Work?

Now let's say that some cost-cutting measures were implemented by this school. The administration found that they could sell one of their buildings if they used their existing buildings more efficiently. Their plan is to shrink the size of each classroom, which they can do if they push all the desks up against each other, with the desks on one side of the classroom pushed right against the wall. This is illustrated in Fig. 4.7. It is clear that there is a significant space savings by taking this approach.

This approach does not come for free, though. Now, when the teacher calls student B3 to the front of the room, all the students in B3's row must get up and walk to the front of the room in a line (Fig. 4.8). There's simply not enough space for them to do it any other way. Once the row is in front, then B3, C3, N3, or any of these students can get to the teacher rapidly. On the other hand, none of these students can reach the front of the room until all the students blocking the way have gone to the front.

This is similar to the way that NAND is laid out. Just as the students share a single aisle at the far side of the classroom, many bits in NAND share a bit line, dramatically reducing the amount of space used on the NAND die to move data back and forth to the bit transistors. Like the classroom analogy, NAND takes a longer time to get a memory bit (a student) that is randomly called to the rest of the system (the front of the class). Once that bit line's data is ready (the student's row is in the front), then data from that word line can be presented to the system at a rapid-fire rate (just as every student in that row can get to the teacher's desk very quickly).

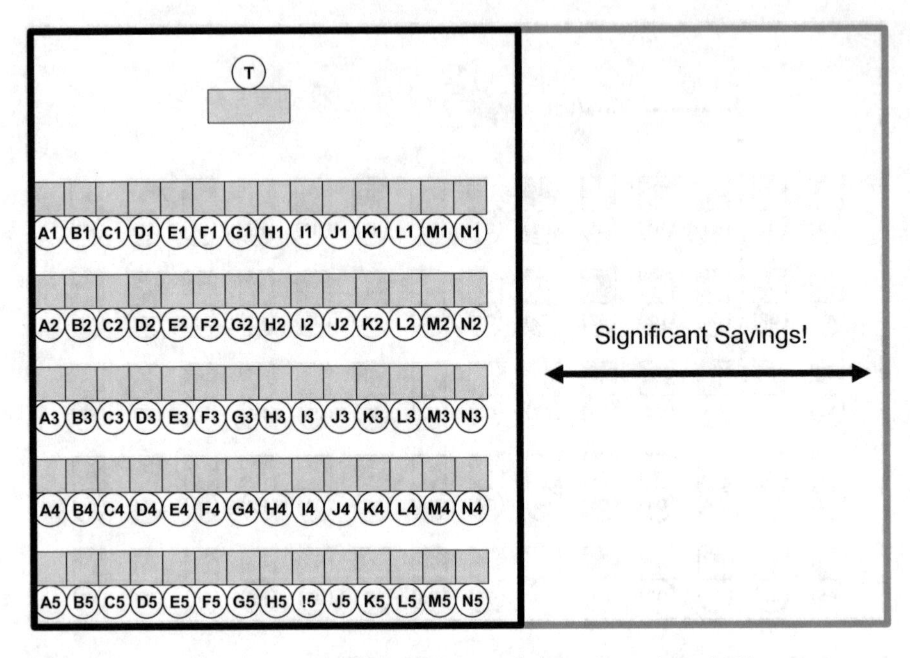

Fig. 4.7 In the NAND classroom, all the desks are pushed together saving valuable space

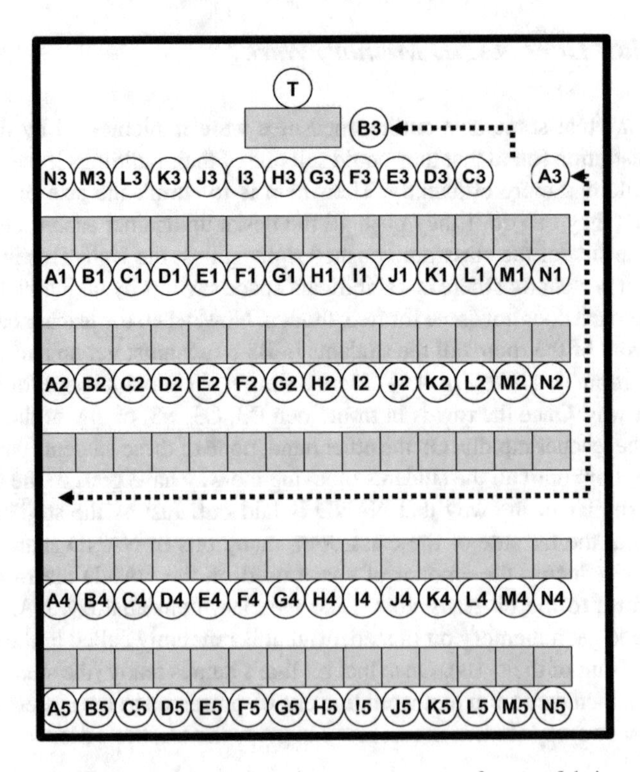

Fig. 4.8 In the NAND class, all the students in a row must move for one of their rowmates to get to the teacher

From this it should be clear that NAND is less expensive to manufacture than NOR, since the manufacturing cost of silicon chip is a function of its size. Keep in mind, though, that the cost savings might be offset by the two technologies' significant difference in functionality. Although NOR usually seems easier to use, surprisingly enough this difference in functionality actually makes NAND a better fit than NOR in applications where serial accesses are preferred to random accesses. Video and audio streaming are two very good examples of applications where serial access is preferable to parallel.

4.6 Bit Errors in NAND Flash

As mentioned earlier, NAND was designed to replace HDDs. The inventors realized that the HDD's innovators had come up with a lot of good ideas and borrowed some of those ideas to further drive cost out of NAND.

One very important similarity is the handling of *bit errors*. The serial technology in NAND is very prone to bit errors, something that digital systems can't tolerate. In order to get past this, NAND makers used error correction technology common to the HDD business, for instance, a 16 K byte page will have an additional 2200 bytes added on. These check or parity bytes are programmed with a code that indicates what the data in those other 16 K bytes should look like.

This implies that every time that a page is programmed, then corresponding *parity bits* must be calculated and stored along with the original bits. A controller is usually given this task. This is one of the reasons that NAND most often ships coupled with a controller chip, although there are cases where the functions of the controller are embedded within another processor within the system.

During a read, the data bits in a page cannot always be trusted, so they are never simply read from the NAND into the system. Instead, the 16 K bytes in the page are processed with the 2200 parity bytes in an error detection and correction engine, which presents a page's worth of corrected data to the system.

Different controller manufacturers use different algorithms in their controllers, resulting in different degrees of correction. You will hear terms like "4-bit correct, 6-bit detect" to describe the capabilities of a controller. The higher the number, the more complex the algorithm required. This means that a more expensive CPU is often necessary to perform that processing in a reasonable period of time. Fortunately, processing power's cost decreases over time, so better and better algorithms are likely to be used in even the lowest-cost systems.

4.7 Managing Wear in NAND and NOR

NAND and NOR memories are treated differently when it comes to managing wear. In many NOR-based systems, no management is used at all, since the NOR is simply used to store code, and data is stored in other devices. In this case, it would take a near-infinite amount of time for wear to become an issue, since the only time the chip would see an erase/write cycle is when the code in the system is being upgraded, which rarely if ever happens over the life of a typical system. In certain other applications where NOR is used to store data, major NOR suppliers provide file management software that implements "wear-leveling" algorithms that will be explained in detail when we discuss NAND below.

NAND is usually found in very different applications than in NOR. Where NOR is most often used for code storage, NAND is most often used for data storage. This means that the end user will continually be erasing and rewriting the data in the NAND. NAND makers manage the wear in the NAND to assure that a cell runs the lowest possible risk of wearing out.

As mentioned above, NAND is always connected to the system by way of a controller. In some cases, this controller is a dedicated chip. Examples of this are SSDs, flash cards, and all USB drives. When you buy one of these, you are buying one or more NAND chips plus a dedicated controller. In other systems, the functions of the controller are implemented within some other processor, perhaps the main processor of the system. In smart phones, for example, the controller is often implemented as a function of the cell phone's baseband processor. One of the responsibilities of this controller is the implementation of the wear-leveling algorithm.

The controller, by being a gatekeeper, can disguise the layout of the NAND chip from the rest of the system. If the system requests the data from address "A," the controller can present a very different address "B" to the NAND and provide this data to the system. This will be OK if the last time the system wrote to "A" the controller forced that data into location "B." The controller keeps track of where everything physically is located and where it looks like it is to the system in a table. This is a relatively simple task.

Another feature of this table is that it tracks how often erase/write cycles have been performed upon each erasable block of the NAND chip. This way the controller can assure that all blocks get an even number of erase/write cycles, thus spreading the wear across the chip. Let's say that a user keeps changing the fifth song in their smart phone's music library but is satisfied with all the rest of the songs. The controller will notice that the space initially used for the fifth song is getting more erase/write activity than other parts of the chip and will decide to move the actual location in the flash memory where the fifth song is stored. If there is free memory, the fifth song will be moved to the free area. If there is no free memory, then something that is rarely changed is likely to get moved to the fifth song's old space, and the fifth song will be moved to that part of the NAND. It is likely in such a case that sections of the NAND are never written to keep moving around, since they sit in prime areas with a very low number of erase/write cycles.

4.8 Bad Block Management

Along with wear leveling, the controller in a NAND-based system is responsible for bad block management. The controller, as a part of its data correction responsibilities, keeps track of how much error correction is performed on each block of the NAND. If the number reaches some predetermined high level, that block is marked as "bad" and will no longer be used by the controller.

This has a very unusual impact upon the user, one that is rarely seen in any digital system: graceful degradation. When most digital systems experience a problem, they come to a sudden stop. The screen goes blank, or the on/off button stops working. In NAND with bad block management, as blocks get taken out of the system, the overall size of the NAND slowly diminishes. In an extreme case, for example, if you had a very small flash card in a system and continually erased photos to snap new ones, you would notice that your card first stored 25 photos, and then 1 day it only stored 24. Later it would slip to 23 and then 22, all very gradually. You wouldn't lose any pictures, but the number of pictures you could store would gradually decline.

4.9 Embedded Versus Removable NAND Flash

Some systems use removable flash memory cards where others have the flash memory soldered into the system. What is the difference between how these systems work?

In reality, the differences between these two kinds of systems are relatively minor. The functions of the controller in a flash card are usually implemented using firmware in an embedded processor. In systems with cards, the main processor of the system is not burdened with error correction, wear leveling, or bad block management, as this is the responsibility of the card's controller. In embedded systems, the wear leveling, bad block management, and error correction are often taken care of by the main processor.

4.10 Flash Memory File Systems

Like an HDD, the flash contains files, and those files must be managed. This is in addition to the error correction, bad block management, and wear leveling discussed above.

This is something that designers need to keep in mind as many firmware designers who start with a small piece of code realize that their system can expand to the point that formal file management software could simplify their data management chores.

File management software is widely available today, and this software has been modeled on software that used for decades by HDDs. In effect, this software hides the difference between the NAND and an HDD for the host device, so that software that was written for NAND could be relatively easily ported to an HDD-based system and vice versa.

4.11 Single-Level Cell and Multilevel Cell Flash Memory

As we saw earlier, the manufacturing cost of a flash chip is a function of the area of the chip. For cost's sake, it is important to get the chip as small as possible. One unique way to do that is to store two, three, or four bits of data in the area where one bit would normally go. This is something that has been employed by hard disk drive and flash makers but has not appeared in other forms of memory.

To understand this concept, you have to look at the basic flash memory cell in Fig. 4.1 as a linear device, rather than as a digital one. The number of electrons stored upon the floating gate is proportional to the time spent charging it, giving the user some control over this charge. The current flowing through the channel will be proportional to the charge on the floating gate when the control gate is not blocking current flow through the transistor.

If you want to store two bits onto a transistor, then the four different states represented by those two bits must each be represented (00, 01, 10, and 11). You can do this by storing four voltage levels. Instead of storing a high charge on the floating gate for a logical 1 and a low charge for a logical 0, using the halfway point as the decision level between a 1 and a 0, you can store four levels: nothing, one third, two thirds, and full, each representing one of the four states. The circuitry for discerning the difference between these levels is a bit more complex, but this technology is well understood.

Chips that store one bit on a cell are usually called SLC for single-level cell chips. Those that store two bits on a cell are called two-level MLC for multilevel cell chips. Chips that store three bits per cell are formally called three-level MLC, but the term TLC (for triple-level cell) is more commonly used. This is a little confusing, since it actually takes eight voltage levels to represent those three bits. Similarly, four-level MLC is often described with a similarly confusing name "QLD" (quad-level cell) even though it is achieved by storing 1 of 16 voltage levels on the bit cell.

The most important issue in making MLC chips is pulling the data out of the device. When a two-level MLC transistor is read, the reading circuit has to discern between four small voltage levels rather than two big ones. With each increase in the number of bits per cell, the size of the voltage difference between the levels shrinks. Digital chips are noisy, and the more noise you have in a system, the more difficult it is to discern small voltages apart from one another. This challenge has caused leading flash suppliers a good share of headaches. One part of the solution is to give the chip time for the noise to settle down before making a decision. This slows the chip down, which is why MLC chips are usually slower on read cycles than their SLC counterparts.

Somewhat similarly, writing into these chips is a bit more daunting than writing into SLC chips. Across the chip, different cells behave differently, and relatively sophisticated state machines (tiny little computers) are used first to put a smidgen of charge onto a floating gate, next to measure to see if the floating gate's charge is near the optimum charge for the cell to represent one of the four levels and then to continue iterating on this process until it is done. Usually a smaller current is used to give better control over the charge, and this slows down programming. Precision takes time.

The biggest issue that must be overcome when putting three or more bits onto a single cell is that the fractional voltages, as they get smaller and smaller, become increasingly difficult to pull out of the noise inherent on a digital chip. Another issue is a mechanism called "adjacent cell disturb." This is when part of the charge on one flash cell leaks sideways into an adjoining cell.

As the levels of charge on the floating gates become smaller and smaller, the impact of a few electrons migrating from one floating gate to another becomes more and more significant. A lot of work has gone into solving these problems and, as a result, there are three- and even four-bit-per-cell products available today. Note that as the number of bits per cell increases, the endurance suffers as well. Higher bit-per-cell NAND flash has become increasingly common, leading to innovations for reducing the wear of the flash memory. With the advent of 3D flash (with thicker lithography and higher raw endurance), three-bit-per-cell SSDs are now very practical, and four-bit-per-cell SSDs are planned.

4.12 Stacking Die to Achieve Higher Storage Capacity

Apart from schemes, to get the most storage on the least amount of silicon is a push to get the most storage in the smallest physical space within the system. This is especially important for portable devices like smart phone handsets and high-capacity flash cards. A common approach to this is to stack memory chips within a plastic package that would normally be used to house a single chip.

Two or more chips are stacked one on top of the other to increase the amount of storage that will fit into the same size package, but it is not uncommon to see stacks with four or eight chips, and there are products that stack as many as 16 chips.

The stacking technology most widely used is to wire-bond chips that have been ground on the back side so they are very thin. Wire bonding is the oldest technology for connecting a die to the outside world, and it is very impressive that it can be done upon multiple levels of chips as shown in Fig. 4.9.

Memory chips are particularly well-suited to stacking, because most systems use memories in such a way that only one chip is powered up at a time, so a hot chip will never be pressed up against another hot chip. This would not be the case with processors or other logic chips like ASICs or *application-specific standard products* (ASSPs).

Of all memory chips, NAND lends itself best to stacking because of its bit errors. This is because memory chips are very sensitive devices, and two chips that are destined for stacking must first be tested to assure that every single bit works flawlessly. Only after they have passed this test can they be stacked and packaged, whereupon they have to be tested all over again and scrapped if they fail.

Although NAND chips are just as prone to damage as are other chips, the fact that the NAND is automatically going to be paired with a controller (that performs error correction and bad block management) allows chips that might otherwise fail to still be useable. In some cases, a semi-functional stack is sold with an unusual storage capacity—some non-binary number like 25GB or 30GB—because of the failures due to the stacking.

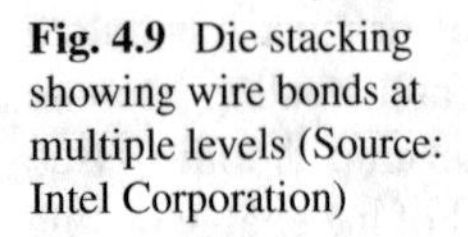

Fig. 4.9 Die stacking showing wire bonds at multiple levels (Source: Intel Corporation)

4.13 Trade-Offs with Multilevel Flash Memory

In the MLC vs. SLC section earlier in this chapter, we mentioned the difference in endurance between SLC NAND and two-level MLC NAND flash. There is a ten-to-one difference in endurance between the two—from 10^5 cycles to 10^4. In today's typical NAND applications, this is unlikely to pose much of a concern. However, as NAND is used to replace other technologies like HDD, and possibly even DRAM in the future, this may be of significant concern.

Say, for example, that a counter that is usually implemented in DRAM is instead implemented in NAND. Just to make the situation worse, let's make this the line counter for a graphics processor: every time the raster scans from one line to the next, the counter is read from memory and then incremented and rewritten.

This is not at all atypical for true DRAM applications, but so far NAND has not been used in this sort of application. One very good reason is that the write speed of flash doesn't accommodate this kind of activity, which for 525 line video would require a new write every 32 μs. This is much less than the 200 μs page write time given in our prior example. Looking past that limitation, though, what would happen if we were able to write to the line counter every 32 μs and the line counter stayed at the same address in the flash?

The simple answer is that 10^5 writes at 32 μs intervals would put the design in jeopardy of a bit failure after its first 3.2 s of use. This is nowhere near acceptable! Keep in mind that this is the SLC endurance. With MLC the system would run the possibility of failing after one tenth that time: 0.32 s!

We choose this example to point out that an engineer needs to consider how to match various memory storage uses to which memory technology, DRAM or NAND.

As was explained in the prior wear-leveling section, the size of the flash in a wear-level-managed system has a big impact upon the time it takes to enter the danger zone of an endurance-related failure. Although two-level MLC's endurance is one tenth that of SLC, MLC's capacity is usually double that of its SLC counterpart, reducing the cost by half. If we assume appropriate use of wear leveling, a designer can afford to throw MLC memory capacity at the problem, rendering the difference between MLC and SLC academic.

As for performance, MLC is slower than SLC, and this point needs to be managed carefully. One approach is to employ parallelism—when one chip is delayed by its slower operation, a second (or third or fourth) chip takes over. In very small systems, this is not a viable approach, but if it is necessary for a system (e.g., an SSD) to use multiple flash chips to achieve its desired storage capacity, there is no reason that these chips cannot be operated in parallel to improve speed.

4.14 Types of Flash Memory Used in CE Devices

Although most consumer electronic devices in the past used NAND in a card format, there is a progression away from that approach as flash content increases.

It appears that the flash content of earlier systems motivated users to store all of one type of data in one flash card and all of another into another card. As one card was filled, its contents would be eventually offloaded to an HDD, while the other was still in use.

Today many CE devices contain such an abundance of flash that this flash can be split into zones where one zone suits one purpose and another suits another purpose, with the threat of the flash becoming filled diminishing over time. This of course assumes that the required capacity of the files also doesn't increase in time.

4.15 Flash Memory Environmental Sensitivity

The key advantage flash memory devices have over electromechanical storage, such as hard disk drives or optical disks, is their extreme ruggedness. From time to time, stories appear attesting to this. SanDisk once issued a press release about a photographer who shot pictures of the demolition of a bridge using a remote-controlled camera. When the camera was struck by flying debris, the lens was shattered, but the flash card remained intact, yielding close-up photos of the explosion.

Another incident involved the fiery crash of a jetliner whose flight recorder was severely burned. Even the epoxy packages on the flash chips were melted. A reverse-engineering firm was able to reattach wires to the flash chip and subsequently read out the data which enabled them to find the cause of the crash.

How much abuse are these chips capable of withstanding?

Standard specifications on a chip of any sort are operating temperatures of –40° to +125 °C, with storage temperatures of up to 150 °C. NAND flash chips follow this convention. The maximum storage temperature is set to one that the package can withstand for extended periods without losing its hermeticity, as moisture is lethal to a chip. The chip inside can withstand temperatures up to about 450 °C before the silicon melts to the point that the impurities start to move around, changing the operation of the device. The epoxy will melt at a far lower temperature of over 300 °C, but the chip will still keep its data after the package is long gone.

Shock and vibration specifications on semiconductors are similar, in that they are specified conservatively to account for weaknesses in the package, rather than in the silicon chip itself. Typical specifications are for high levels of constant acceleration that is used to test if the bonding wires will pull off the chip, which once again is a test of the package rather than of the silicon itself.

4.16 Using Memory Reliability Specifications to Estimate Product Lifetime

There are few specifications that must be understood to estimate the likelihood of failure in a flash chip. This is particularly important to understand if the engineer wants to design a controller rather than to purchase one from a controller vendor.

The most important specification is *endurance*. This is a measure of how many erase/write cycles can be performed before a bit failure is likely to occur. The guaranteed minimum for most makers of SLC flash is 10^5 or one hundred thousand erase/write cycles. For two-level MLC, the guaranteed minimum ranges from 10^3 to 10^4 cycles, and for three-level MLC, endurance can be specified as low as 300 cycles.

The next most important reliability specification is *data retention*, which is the manufacturer's guarantee of how long bits will remain reliable if they are never rewritten in a worst-case environment. Ten years is the norm here, which is usually longer than the life of the device that is using the flash. There is a complex relationship between endurance and retention, though. The more wear a flash chip experiences, the lower its retention will be. Each chip maker's specification is different, though, so it is best to have a close look at the specifications for individual chips if long data retention is required in a high-wear application.

In order to really understand endurance, an engineer must know the erase and programming specifications of the device. Flash chips (NAND and NOR) are very sluggish to erase compared to other memory types, and erase cycles tend to be in the 1.5–2 ms range. These erases clear an entire block. Each block is composed of pages, and a page programming time of 200 μs is typical.

Other specifications required to understand these specifications are the block size (16 K bytes in a small-block NAND and 128 K bytes in a large-block NAND) and the page size (typically 16 K bytes). Erases are performed on blocks, and writes are performed on pages. These specifications with the overall size of the NAND chip will give the engineer enough to calculate a worst-case scenario for an endurance failure.

4.17 Flash Memory Cell Lifetimes and Wear-Leveling Algorithms

The real limiting factor that stands in the way of wear-out is the long time it takes to erase and reprogram a block. A typical block erase takes 1.5 ms and another 200 μs is needed to program a page within that block. To perform these two functions, 10^4

times would take about 23 min. This assumes that the system spends all of its time repeatedly erasing and programming a single block in an attempt to cause it to fail.

Let's now look at a 256Gb chip, which was a typical size for an MLC NAND at the writing of this chapter. Such a device has four planes of 548 blocks each for a total of 2192 blocks. If a system spent all of its time erase/programming blocks in a round-robin order to assure that all were hit equally, and if only one page per block was programmed (the very worst case), a failure might occur in as soon as 36 days. More realistically, if every page were programmed in a block before moving to the next block, since this part has 1024 pages per block, the exercise would consume 582 days or about 1½ years.

Remember that this example assumes that the device is never being read from; it is simply undergoing constant erase/writes. A few systems might behave this way, for example, a black box recorder, but only if the bandwidth needs were rather extreme. Otherwise there would most likely be breaks between these cycles, prolonging the time before a failure. If a system spent a certain portion of its time idle, another portion reading, and another portion performing erase/writes, the 1½ years just calculated would most likely stretch out for a number of years.

Also recall that the sole consequence of a bit failure will be that a block will be disabled, effectively shrinking the size of the chip. There will be no catastrophic failure. The NAND will not have to be replaced unless every last block is disabled.

This is why wear leveling works so well. The difference between the first example's 23 min and the 1½ years case attests to what can be accomplished if the controller is smart enough to track every block's activity. To look at this another way, wear leveling in a chip with 1024 pages per block and 2192 blocks per chip turns the 10^5 erase/write cycles into an effective 10^{11} cycles, getting an improvement of 100,000 times. Note that this 10^{11} refers to the number of page writes that can be performed into the chip before a single bit error is likely to occur if the pages are being written into in sequence. Assuming that 1 MB photo files are being written into the chip this way (consuming 8 pages per photo) nearly three billion photos would have to be cycled through the part before a bit failure would be likely to occur.

Wear leveling spreads out the problems of endurance or wears across the entire chip, so the bigger the chip the less frequently a wear failure will strike. With each doubling in a chip's density, the number of photos or the time between failures will also double. One company has posed this as an argument for decreasing endurance specifications. Since the incidence of bit failures halves for every doubling of chip density, does it make sense for a 256Gb chip to have to exhibit the same endurance as its 128Gb predecessor? It is likely that there are chips today that quietly relax this specification, counting on sheer size and wear leveling to hide this change from the user's eyes.

Assuming that the chip is jam-packed with information with no free space at all, each update will require data to be exchanged between a relatively inactive block and a more active one, when the system tries to write into the more active block. This means that each erase/write will result in two erase/writes to manage wear leveling. Although this could cut the endurance of the chip in half, this unlikely scenario still would hit a limit of 1.5 billion photos or about 4 years of nonstop writing.

4.18 Predicting NAND Bit Errors Based upon Worst-Case Usage

The above analyses have been performed only for the extreme case. For real-life evaluations, a more practical approach is in order. The designer needs to either understand usage patterns for actual analysis or to make assumptions (which will need to be fully disclosed to the boss) that will provide a reasonable usage scenario.

Let's take an example of a music library on a smart phone. We will assume that 16GB of storage is devoted to this music library and it can store 4000 songs. The most trying user might download an entire new collection of 4000 songs once a day.

If we choose a two-level MLC NAND (typical) with a 10^4 endurance, and assume that no bit errors can be tolerated by the controller (unlikely), the simplest equation says that the entire music library can be reprogrammed 10^4 times before a bit error is likely to be encountered. This means that it could fail after 10^4 days or about 27 years. If a smaller number of songs is stored, and the controller performs wear leveling, those 27 years would be extended by the inverse of the fraction of the memory that is used (i.e., if ¾ of the memory is used, then the endurance would extend to 4/3 times 27 years or about 36 years). Of course, this is all academic because a consumer electronic device is unlikely to be used for even 5 years before being replaced.

Let's evaluate another scenario. Although it is uncommon for an amateur photographer to spend the entire day shooting photos every day of the year, it is conceivable that a professional photographer would take one photo every 5 s for an 8 h workday every workday of the year. This would amount to about 1.4 million photos per year. At this rate, if the photographer constantly used the same very small 1GB card to shoot 2 MB photos that card would have a bit error about once every 2 years.

We should note that a 1GB card is extremely small for today's professionals, especially. At the 5-s rate, the photographer would be shooting 720 photos per hour, filling the card relatively frequently, so time would have to be taken simply to read the photos off the card or to erase them in the camera. It is most likely that this photographer would opt for a significantly larger card, which would extend the lifetime of the card in proportion to its size—a card twice as large would last twice as long. Purists shoot in raw format, which might expand that 2 MB photo size used in the preceding analysis to about 45 MB. This would get the photographer only 22 photos on that card 1GB card or less than two min worth of shooting.

As pointed out before, the frequency of failures declines as the card size increases, so this particular usage pattern would most likely require the photographer to suffer from a bit failure every several years.

Both of these relatively extreme users see a far longer time between failures than the 1½ years outlined in the previous worst-case scenarios. This illustrates the need to understand the usage expected in the system to understand expected failure rates.

4.19 Flash Memory Format Specifications and Characteristics

There are numerous flash card formats. This section will give a brief overview of the predominant types.

4.19.1 CompactFlash (CF) and Related Card Formats

The CF format is 3.3 mm thick and 36.3 × 43.8 mm in length and width. A 5.0 mm thickness is also specified, but this specification appears never to have been used for NAND flash, only for small microdrive HDDs.

The CF format uses a 50 pin socket (as opposed to the card-edge connectors used by all other card formats) and communicates through an ATA interface that supports capacities up to 137GB. Data transfer rates have been upgraded from 8 MB/sec to 66 MB/sec and could be further enhanced in the future.

In 2008, a variant of CompactFlash, CFast, was announced. CFast (also known as CompactFast) is based on the serial ATA interface. In November 2010, SanDisk, Sony, and Nikon presented a next-generation card format to the CompactFlash Association. The new format has a similar form factor to CF/CFast but is based on the PCI Express interface instead of parallel ATA or serial ATA. With potential read and write speeds of 1 Gbit/s (125 MByte/s) and storage capabilities beyond 2 TB, the new format is aimed at high-definition camcorders and high-resolution digital cameras. These XQD cards are not backward compatible with either CompactFlash or CFast. The XQD card format was officially announced by the CompactFlash Association in December 2011.

4.19.2 Multimedia Cards (MMC)

SanDisk and Siemens introduced the MMC in 1997. This format is 32 × 24 × 1.4 mm and uses a card-edge connector to save space and cost.

This format has undergone a series of evolutionary changes, shrinking in size to support the MMC*mobile* format at 24 × 18 × 1.4 mm and then the MMC*micro* at 12 × 14 × 1 mm. Another evolutionary change has been the communication speed: read speed has been increased from 2.5 MB/s to 52 MB/s over time, while write speed has increased from 200 KB/s to as high as 10 MB/s, limited only by the speed of the NAND chips within the card. The maximum capacity supported by the MMC format is 2 TB (as of 2007).

4.19.3 Secure Digital (SD) Cards

Shortly after the introduction of the MMC, the Secure Digital or SD card was introduced. This card has the same mechanical format as the MMC and uses a somewhat compatible interface (MMC cards can be used in an SD slot, but not vice versa) but includes a more sophisticated controller that addressed the security concerns of the media business, who wanted some assurances that media would not be replicated with abandon by the users of these cards. The MMC specification was upgraded after the introduction of the SD card to include digital rights management features.

As card formats shrank, so did the SD card, moving to the miniSD at 21.5 × 20 × 1.4 mm, followed by the microSD at 11 × 15 × 1 mm. The normal SDHC card interface tops out at a 12.5 MB/s communication rate, and the format supports capacities of up to 32GB. The SDXC standard allows 32GB to 2 TB capacity. Ultrahigh-speed (UHS) SDHC and SDXC cards have data rates as high as 312 MB/s.

Table 4.1 below compares the different card formats in terms of volume, capacity, and read speed.

4.20 Flash Memory and Other Solid-State Storage Technology Development

4.20.1 Road Map for Flash Memory Development

The driving factor for flash card and USB flash drive capacity increases is the increase in capacity of NAND flash chips. NAND chips account for as much as 95% of the manufacturing cost of a flash card or USB flash drive. The increase in NAND chip capacity drives not only the amount of flash that can be fit into one of these card or USB formats, but it also drives the affordability of that amount of storage. At the writing of this book, a 64 gigabyte USB flash drive is very affordable at under $20, whereas only 2 years earlier, it would have cost twice that amount. Future readers are likely to laugh at how expensive these numbers seem.

Table 4.1 Key attributes of leading flash card formats

Format	Volume	Max capacity	Read speed
CompactFlash	5.1 cm^3	137 GB	132 MB/s
CFast	5.1 cm^3	>512 GB	600 MB/s
XQD	5.1 cm^3	2 TB	1 GB/s
MMC	1.1 cm^3	4 GB	52 MB/s
MMC*mobile*	0.6 cm^3	4 GB	52 MB/s
MMC*micro*	0.2 cm^3	4 GB	52 MB/s
SD card	1.6 cm^3	2 TB	312 MB/s
miniSD	0.6 cm^3	2 TB	312 MB/s
microSD	0.2 cm^3	2 TB	312 MB/s

NAND chip vendors drive their costs down and their capacities up by shrinking the minimum size feature they can image onto a piece of silicon at a continuing progression. At the writing of this book, most NAND flash vendors are migrating their production capacity from processes with minimum feature sizes of 20 nm (nanometers or millionths of millimeters) and 15 nm to 3D processes. Many applications currently use three bits per cell and are expected to migrate to three bits per cell to triple or quadruple the capacity of an SLC flash device.

The cost of a planar NAND flash chip is inversely proportional to the square of its process geometry, that is, a 128Gb (gigabit) chip produced on a 15 nm process would be $15^2/20^2$ or 56% the size of a 128Gb chip produced on a 20 nm process. This implies that the cost of a 15 nm 128Gb chip should be about 56% that of its 20 nm counterpart. As the process evolves, the capacity that can fit onto a card is likely to go up, and the price for that much capacity has room to go down.

With 3D NAND the cost is roughly proportional to the number of layers of bits: a 64-layer NAND flash chip should cost a little more than half as much as the same capacity chip made using a 32-layer process.

4.20.2 *Expected Growth in Storage Capacity for Flash Memory*

There is a steady upward progression in NAND chip capacity and a steady reduction in the price per GB. This trend grows the amount of storage that can be sold at certain price points and how much storage can be fitted into a given form factor.

This evolution drives the chart in Fig. 4.10, which illustrates the likely average capacity of a NAND flash chip for the next 3 years and the historical perspective for the past 7 years.

The chart uses logarithmic y-axes, since the use of a linear y-axis makes all but the extreme ends of the chart become very difficult to read.

Beyond 15 nm the progression of planar NAND reached a limit, since planar NAND flash cannot successfully be made to operate past the 15 nm node. This limit has driven NAND flash makers to adopt an alternative technology to continue to push NAND's cost structure farther than was possible using the current planar technology.

The issue stemmed in part from the ever-diminishing number of electrons used to store a bit. This drove the market to convert from planar NAND flash to a different technology: 3D NAND flash. As shown in Fig. 4.11, a 3D (three-dimensional) NAND stacks bit cells rather than shrinking them. The number of bits per chip increases in proportion to the number of layers even though the process is not scaled the way that it has been in the past. In practice, this has allowed the flash memory manufacturers to relax on the lithographic requirements per cell and thus improve endurance and data retention for a give number of bits per cell. A 3D flash requires a more complex manufacturing process, and although all the flash memory companies are moving to 3D flash, it has proven challenging to achieve cost parity with planar flash memory. Before 2020 3D flash will likely replace almost all planar flash.

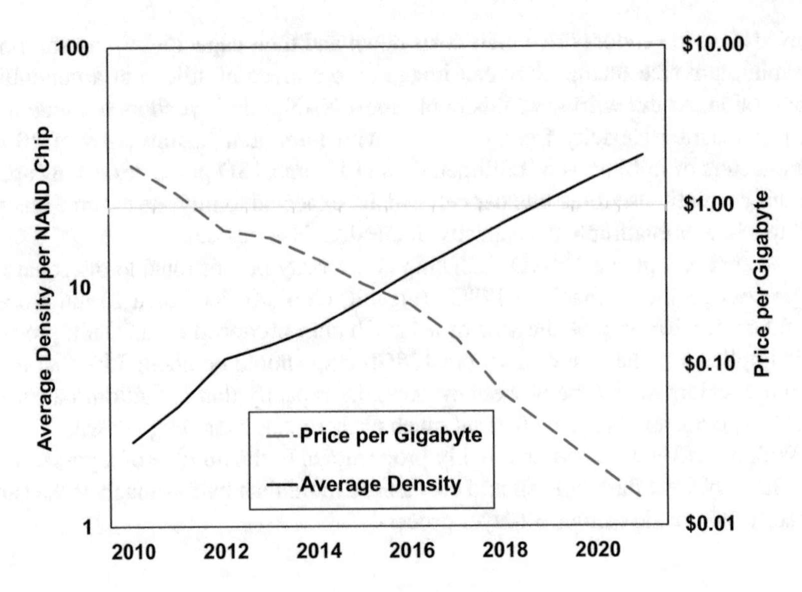

Fig. 4.10 NAND average chip density and price per gigabyte trends

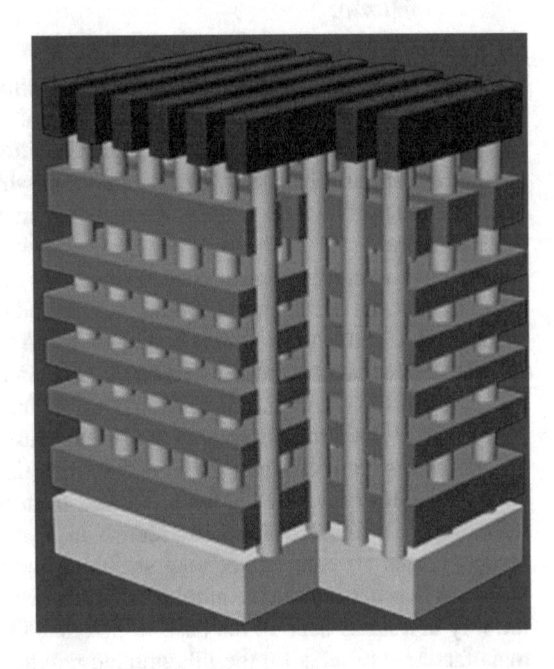

Fig. 4.11 Toshiba's BiCS vertical NAND structure (Courtesy of Toshiba Corp)

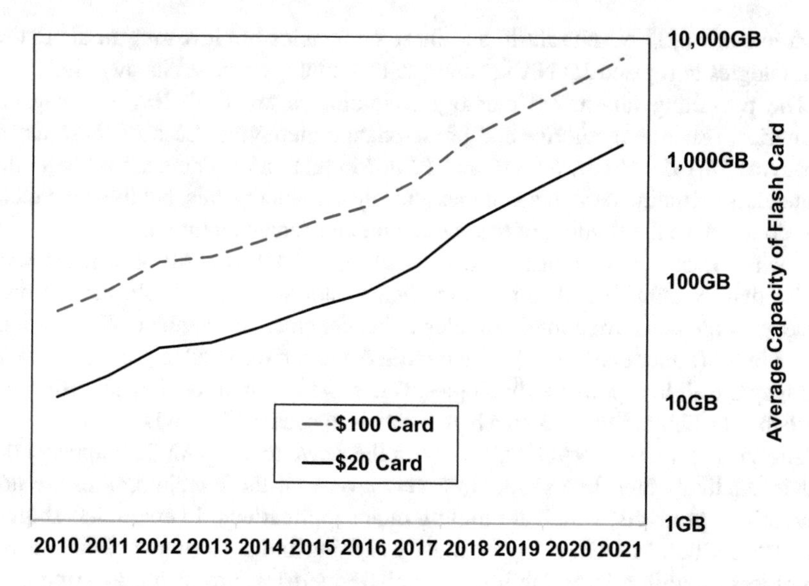

Fig. 4.12 Growth trends for $20 and $100 flash cards

4.20.3 Expected Change in Cost per GB of Flash Memory

As was mentioned before, the major cost component in flash cards and USB flash drives (and any other flash-based storage medium) is the cost of the NAND flash chips. It is only reasonable to assume, then, that the costs of flash cards are likely to follow this in some way. It is unlikely, though, that flash cards will go from $20 today to $10 next year, and then $5 the following year, falling below $1 and onward as the price for a gigabyte of flash falls. Instead, the price of cards is likely to follow certain consumer price points (say $20 and $100) and that the capacity of these cards will rise to meet these price points.

Figure 4.12 is an illustration of how that is likely to happen. In this chart, we have taken the prices used in Fig. 4.10 to determine the average capacity of cards priced at $20 and $100 over the same time frame. Certain assumptions have been made in this chart about other costs and about retail markup, but the trend is clear; the average capacity of a $20 card is likely to double every year as costs decline by a similar rate.

4.20.4 Other Solid-State Storage Technologies

There was some mention above about the difficulties faced by flash designers as manufacturing processes continue to shrink. A roadblock to continued process migration has already been encountered that forced flash makers to migrate from planar to 3D NAND flash, and these companies are investing in alternative technologies to replace 3D NAND once that technology reaches its own limits.

The prevailing alternative memory technologies are ferroelectrics, magnetic memories, resistive memories and phase-change memories. Each of these has an acronym: *FRAM*, *MRAM*, *RRAM*, and *PRAM* in that order. There are several other contenders including carbon nanotubes and silicon nanocrystals, but these technologies do not have the funding of the leading memory manufacturers.

Each of these technologies has certain strong advantages. All offer faster write cycles than NAND, and all offer much better endurance than flash. Furthermore, these technologies have a much simpler write and erase mechanism, allowing individual bytes (rather than pages) to be modified without the need to pre-erase blocks, and they are all low-power technologies. But, most importantly, they all promise to scale to ever-tightening processes better than is expected of NAND.

Are these threats to NAND flash? With the move to 3D NAND, it appears that this is not likely any time soon. However, several of these technologies are now available in products, which are finding niche applications. Everspin has shipped over 100 million MRAM chips for mostly caching and buffering applications and a partnership with a large fab house, GlobalFoundries, promising to bring spin-tunnel torque MRAM to embedded applications. STT-MRAM is nearly as fast as DRAM and just about as durable, but it is nonvolatile. Thus, it may replace DRAM as the price drops, with higher volume manufacturing. Replacing volatile DRAM that must be regularly refreshed with a nonvolatile memory offers significant power savings by avoiding the refreshes and also allowing power saving modes, since the nonvolatile memory will have instant on. Besides Everspin, there are a few start-ups and many large companies working on their versions of MRAM technology.

Phase-change memory is behind the 3D XPoint technology that Intel and Micron introduced in 2015. The companies have stated that they want 3D XPoint technology to be used between NAND flash and DRAM. Thus, this technology could offer an intermediate and nonvolatile technology that reduces the amount of expensive DRAM needed in a hybrid memory system. It is possible that this new storage technology will also find its ways into consumer applications as volume ramps up.

Several companies have plans to offer various types of resistive RAM (RRAM) that is not a phase-change memory, and some have speculated that RRAM could replace NAND flash when 3D flash memory runs out of gas. However, it now appears that this will not happen until sometime in the 2020s at the earliest. RRAM will need to find market niches where it can offer advantages if it hopes to survive until flash memory dies.

It is reasonable to say that NAND will be replaced by one of these technologies before this book disintegrates from age, but it does not appear reasonable to assume that they will gain much ground over the next 5 years.

4.21 Chapter Summary

- Flash memory technology is an application of the floating gate or charge trap semiconductor technology. Flash memory is a nonvolatile storage technology that is widely used in mobile consumer electronic applications.
- The basic operations of a flash memory device are reading, writing, and erasing. In order to rewrite data on a flash memory device, the flash memory cells must first undergo an erasure step.
- Over repeated use, the insulating tunneling layer in a flash memory device can accumulate charges from electrons that do not tunnel through. As the number of trapped charges increases, it becomes difficult to keep a charge in the floating gate or charge trap. This accumulation of charge is referred to as "wear" of the flash memory cell and eventually leads to the inability to maintain data on that cell.
- There are two kinds of flash memory, NOR and NAND. The two terms are names of types of logic gates, the negated "or" function and the negated "and" function. The difference between the two types of architectures is real estate. NAND has a significantly smaller die size than does NOR. This translates to significant cost savings. NAND is different from other memory technologies in that it is partially serial and partially parallel.
- The serial technology in NAND is very prone to bit errors. Like hard disk drives, NAND makers used error correction technology to correct bit errors. A page of data will have some parity bits added to check for errors in the page data. These check or parity bits are programmed with a code that indicates what the data in those other 16 K bytes should look like.
- Wear leveling is a technique to spread the writing and erasing over all the cells in the flash memory so that individual cells do not wear out before the rest of the memory. Wear leveling is managed by the flash memory controller chip.
- The controller is also responsible for bad block management in the memory. The controller keeps track of the number of bad bits in each block, and if the number of bad bits becomes too large, the block is declared bad and not used any more. The data that was on that block is then moved to an unused "spare" block.
- Flash memory can be used as a chip installed within the circuit board of a device or it may be embedded in a removable card.
- Like an HDD, the flash contains files, and those files must be managed. A flash file system makes the data on the flash memory device look like it comes from a hard disk drive to the host device.
- In order to increase the storage capacity for a NAND flash memory device with a fixed number of memory cells, the number of bits per cell can be increased. Increasing the number of bits per cell involves splitting the voltage levels that are used to read the bits on the cell. Multilevel cells (MLCs) trade off increasing storage capacity for greater sensitivity to flash wear during the erase and write cycles. Also, the noise on the cells does not decrease as the cell voltages are split, and so there are more errors that need correction with a MLC flash memory. Implementation of MLC or even higher number of bits per cell

involves more sophisticated controller technology and this sophistication, and processing power has to increase as the number of bits per cell increases.

- Flash memory die can be stacked on top of each other to achieve greater volumetric storage capacity. Flash memories are particularly good at this since they generate lower operating power (that needs to be dissipated) vs. other types of semiconductor chips.
- Flash reliability is dependent upon careful control of cell wear and error correction. Since these operations occur in the flash memory controller, controller technology is key to providing specified reliability. This is even more important with MLC flash. Flash memory is inherently insensitive to shock or vibration since it is a solid-state storage technology. Used in the right way and in the right applications, flash memory can give many years of useful life in a consumer product.
- Removable flash memory card formats have developed over the years due to new applications, form factors, and proprietary designs. Today the most common flash card formats are SD cards, MMC cards, and miniaturized versions of these formats.
- Over the last few years, flash memory has undergone a significant reduction in price for given storage capacity points. This price reduction expressed in $/GB should continue at about a 30% average annual rate for the near future (with declines in any year driven by many factors). As affordable flash memory storage capacity increases, the number of applications that flash memory can participate in increases. With the lower base price for flash memory (compared to other technologies such as hard disk drives), flash memory will be found in lower storage capacity CE products with growth potential into higher-capacity products.
- 3D flash has allowed the use of relaxed lithographic features, allowing higher-capacity flash memory with higher endurance and up to four bits per cell. 3D flash memory will replace planar flash in the next few years and has many generations of higher-capacity products for many future generations.
- Emerging memory technologies like MRAM, FRAM, and PCRAM could fill a niche between flash and DRAM or even replace DRAM. Sometime in the future, one of these technologies might replace flash, but not soon.

Chapter 5
Storage in Home Consumer Electronic Devices

5.1 Objectives in this Chapter

- Look at the functions, design trade-offs, and uses of digital storage in common household consumer electronic products.
- Consider the development of digital storage in digital video recorders and how this impacts direct attached and networked home storage versus storage in the cloud.
- Compare in-home network storage for DVRs to network DVR initiatives to consolidate digital storage needs, particularly for service providers.
- Learn about IP set-top boxes and Smart TVs
- Examine the trends in digital storage for game consoles.
- Examine how home storage networking could expand as part of a smart connected home to create a home storage utility.
- Look at video requirements in the future and their impact on digital storage capacity requirements in the home.

Digital storage is an enabling technology for many modern consumer devices. This includes products such as digital video recorders (DVRs), home media centers, home network storage, automobile entertainment and navigation systems, mobile music and video players, digital still cameras, camcorders, cell phones, game consoles, and many other devices. Digital storage allows the use of rich content in these devices as well as allowing personal content creation. This chapter and the next explore many of these devices used in the home and when we are on the go. These chapters show how digital storage is used, the important characteristics for digital storage used in these applications, and finally where each of these applications could be heading in the future. We will pay attention to how digital storage will impact the development of these devices.

T.M. Coughlin, *Digital Storage in Consumer Electronics*,
https://doi.org/10.1007/978-3-319-69907-3_5

5.2 Personal Video Recorders or Digital Video Recorders

Digital video recorders (DVRs also known as personal video recorders or PVRs) have dramatically changed customer viewing habits. A digital video recorder can record and play back full motion video and audio signals using the ISO *MPEG* audio/video standards for the compression of audio and video content. This differs from the analog *video cassette recorders* (VCRs) that were the first commercial devices to free consumers from a fixed television program schedule. Storage in digital video recorders is usually done with hard disk drives.

There is a long history of using HDDs to record and play back video. Ampex introduced a hard disk analog video recorder in 1967, the HS-100. The HS-100 was developed to give CBS instant freeze-frame capability. The product recorded composite analog video onto a 14″ diameter hard disk using FM modulation. The total storage time was only 30 s, but the device could record continuously and play back from twice the normal speed down to still frame.

In 1985, an employee of Honeywell's Physical Sciences Center, David Rafner, first described a drive-based DVR designed for home TV recording, time-slipping, and commercial skipping. The [US patent 4,972,396] focused on a multichannel design to allow simultaneous independent recording and playback. Broadly anticipating future DVR developments, it describes possible applications such as streaming compression, editing, captioning, multichannel security monitoring, military sensor platforms, and remotely piloted vehicles.[1]

TiVo and ReplayTV launched their digital video recorder (DVR) products at the 1999 Consumer Electronics Show. Since their introduction, DVRs have revolutionized home video viewing habits in a way that VHS-based home video recorders could not. By using digital recording technologies, it became easier to access particular content and store it on hard disk drives. With the introduction of *electronic program guides* for these products, they could be set to automatically obtain video content from a cable or satellite network source and store it for later viewing. This intelligent time shifting has forever changed consumer viewing habits and threatened traditional advertising models since customers have been able to skip ahead or fast-forward through undesired advertising.

Besides stand-alone DVRs produced primarily by TiVo in the USA (and vendors such as Channel Master, Humax, Nuvyyo, Thomson, TiVo, Topfield, Pace, and Vestel elsewhere in the world), this technology is often embedded in modern cable or satellite *set-top boxes* where such features have helped prevent customer churn. They are also sometimes bundled into DVD recorders which can be used for archiving video content. This last feature is particularly popular in Asia. Companies that manufacture set-top boxes (STBs) with (or without) DVR capability include ARRIS, Huawei, Pace, Technicolor, and several large Chinese manufacturers.

DVR capability has also been incorporated into home media center PCs and even into some TVs by companies such as LG and Hitachi, although Smart TVs generally support streaming services rather than recording. Video recording capability is very popular and this function is often used on mobile devices as well.

[1] Wikipedia article on digital video recorders, http://en.wikipedia.org/wiki/Digital_video_recorder

Fig. 5.1 TiVo Series 2 digital video recorder

Fig. 5.2 Interior of TiVo Series 2 digital video recorder

5.2.1 Basic Layout and Design of Digital Video Recorders

Figures 5.1, 5.2 and 5.3 show a DVR product manufactured by TiVo (one of the founders of this business).[2] Figure 5.1 shows an example of digital video recorder set to record up to 80 hours of standard definition TV. This product came with an 80 GB internal hard disk drive manufactured by Maxtor (now part of Seagate Technology). Series 2 TiVos like these include a USB port (for Ethernet and 802.11 adapters), a faster CPU than the first TiVo units, and more RAM. TiVo took advantage of this network connectivity with software features like TiVo ToGo and Home Media Engine applications.

Figure 5.2 shows the interior of an early TiVo box. 3.5 inch hard disk drives are used in these products since they provide the highest capacity at the lowest price. As can be seen, there is a lot of empty space in this box. Current DVR (PVR) products are much more efficient in their use of space. Figure 5.3 shows a close-up of the main circuit board of this device identifying several of the ICs on the board.

[2] Inside Peek report on TiVo Digital Video Recorder, 2007, http://www.inside-peek.com/

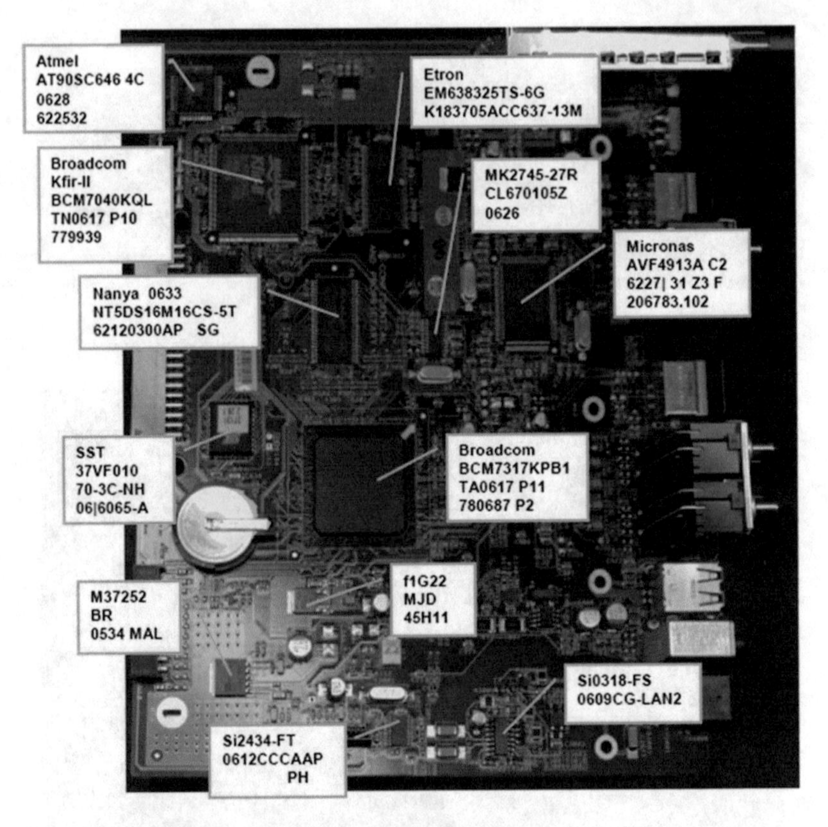

Fig. 5.3 Main circuit board of TiVo Series 2 DVR showing many of the ICs

Figure 5.4 shows a reference design for a digital video recorder from Texas Instruments.[3] The DVR block diagram includes several components (with many functions incorporated into the master chip):

- The device includes a 1 GHz ARM Cortex-A8, 1 GHz TI C674x floating-point DSP, several second-generation programmable HD video image coprocessors, and an HD video processing.
- The system utilizes comprehensive codec support including H.264, MPEG-4, and VC1 at HD resolutions.
- It simultaneously supports three channels of 1080p at 60 fps.
- Glass-to-glass latency is reduced below 50 ms.
- Multiple interfaces include gigabit Ethernet, PCI Express, SATA2, DDR2, DDR3, USB 2.0, MMC/SD, HDMI, and DVI.
- It seamlessly capture up to 16 D1 video channels.

[3] Texas Instruments, DM8168 Based DVR System Block Diagram, TI Reference Design Library.

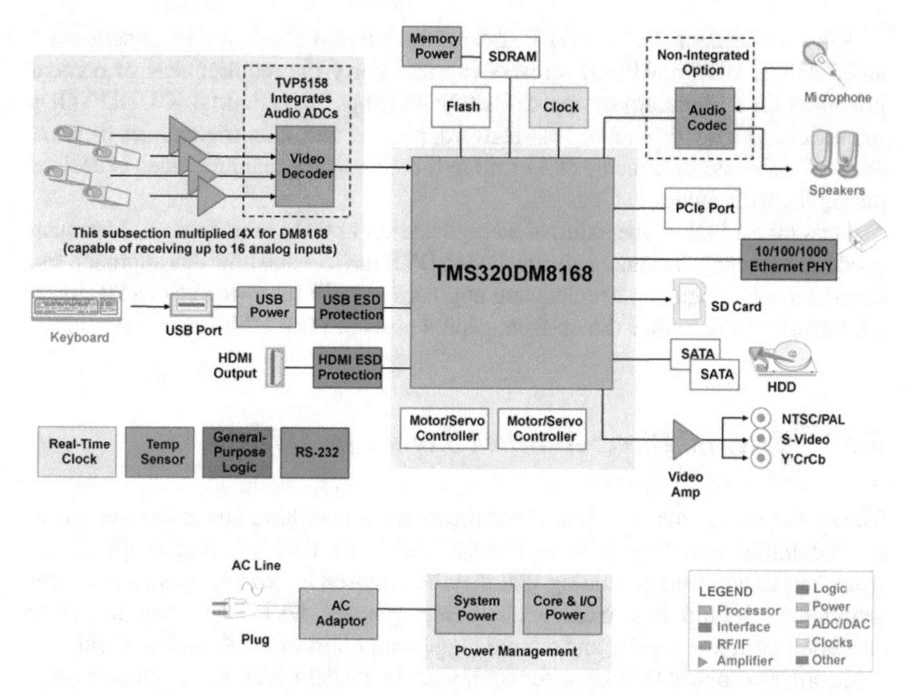

Fig. 5.4 Reference design block diagram for digital video recorder (Courtesy Texas Instruments)

5.2.2 Digital Video Storage Requirements and DVR Storage Design

Richer video content requires higher-capacity storage as well as faster data rates. For instance, an SD video requires about 1.25 GB per hour of storage, while an HD video requires about 5.0 GB per hour. A 4 K UHD video requires at least 20 GB per hour (depending upon the compression being used). Thus a 500 GB disk drive can hold about 400 h of SD, 100 h of HD content, and about 20 hours of 4 K UDH content. An HD or UHD recorder requires access to larger data storage. This requires the use of larger capacity hard disk drives with these products. Such greater storage capacity could be provided internally with an embedded hard disk drive. It could also be available through a *direct-attached storage* device (DAS or non-shared storage) at the back of the DVR or through a network connection on the DVR attaching the device to network-attached storage.

In addition to storage requirements, the data rate required by the device for storage and any network and external interfaces depends upon the resolution of the content and the desired special or trick features. The vast majority of broadcast content in the USA today is either MPEG 2 or MGEG 4 (for UHD content). Measurements of streams from the top four US service providers show that SDTV MPEG2 broadcasts are variable bit rate with average data rates between 2 and 3 Mbps, while peak rates are around 9 Mbps.

Video on demand (VOD) SDTV streams tend to be a constant bit rate between 4 and 5 Mbps. HDTV MPEG2 streams can also carry the requirements of content providers for a minimum of approximately 12 Mbps allocation.[4] 4 K UHD VODs require even higher data rates. Fast-forward, reverse, and other trick modes increase the peak data rate by a factor of 3 or more if continuous streaming video is desired during the trick mode.

Let's take a look at the external storage add-on options to understand how such products are being designed and attached to DVR devices and how this approach can be used to avoid field replacement and upgrades of DVR in set-top box (STB) internal storage and to reduce the up-front capital costs of DVR STB implementation.

5.2.3 External Direct-Attached Storage for DVRs

The earlier generation of digital video recorders did not have any active options to expand digital storage capacity beyond that built inside the DVR. As a consequence, users had to erase programming that they have stored in order to record new programming. Set-top box manufacturers incorporated DVR capability to create increased customer loyalty and are the biggest suppliers of DVRs to the world.

In order to control their costs for equipment in the field, STB manufacturers want to ship product using the current volume disk drive capacity rather than higher-capacity drives that cost considerably more. Thus, most DVRs do not have very high-capacity hard disk drives nor is there much incentive to supply higher-capacity disk drives in STB DVRs. One way to provide extra capacity is to attach an external storage device to expand the overall capacity in the DVR.

Table 5.1 shows technical comparisons between various external and internal storage interfaces showing differences in data rate, cable length, power on the line, and devices per channel. The fastest interfaces are the USB and Thunderbolt interfaces. USB is the most common storage interfaces in consumer products. Firewire (IEEE 1394) is an older interface that was primarily used as a video recorder connection interface to move content from a camera to a computer video editing workstation. eSATA is an interface technology based upon the serial ATA (SATA) internal disk drive interface that allows very fast data transfers (up to 6 Gbps) and also multiple device connection using port multiplication. There are several set-top boxes that include eSATA interfaces for external expansion storage.

SATA interface disk drives became available in 2005, and in 2007, they became predominate over the older parallel ATA interface. There are no parallel ATA interface hard disk drives produced today. Current SATA and eSATA products have 8b/10b encoded data rates of 16 and 6 Gbps, respectively.

Consumers would like to have a greater capability to retain their DVR content, and an external drive expansion option is very attractive to DVR customers. A USB

[4] Leveraging MOCA for In Home Networking, Anton Monk, http://www.convergedigest.com/bp-ttp/bp3.asp?ID=453&ctgy=Home

Table 5.1 Comparison of internal and external storage interfaces

	eSATA Rev. 3 (external)	SATA 3.3 (internal)	FireWire 800 (external)	USB 2.0 (external)	USB 3.0 (external)	USB 3.1 (external)	Thunderbolt 2.0 (external)	Thunderbolt 3.0 (external)
Data rate (Mbps)	6000	16,000	786	480 (burst)	5000 (burst)	10,000 (burst)	20,000 (bidirectional)	40,000 (bidirectional)
Max. cable length (m)	2	1	4.5 (16 cables can bedaisy chained up to 72 m)	5 (USB hubs can be daisy chained up to 25 m)	3 (USB hubs can be daisy chained up to 15 m)	3 (USB hubs can be daisy chained up to 15 m)	3 (Cu), 60 (optical)	2 (Cu), 60 (optical) USB-C
Power provided	No	No	15 W, 1.2A	2.5 W, 500 mA	4.5 W, 900 mA	100 W, 5A	No	100 W, 5A

Fig. 5.5 USB expansion of a DVR allows only a single external storage device

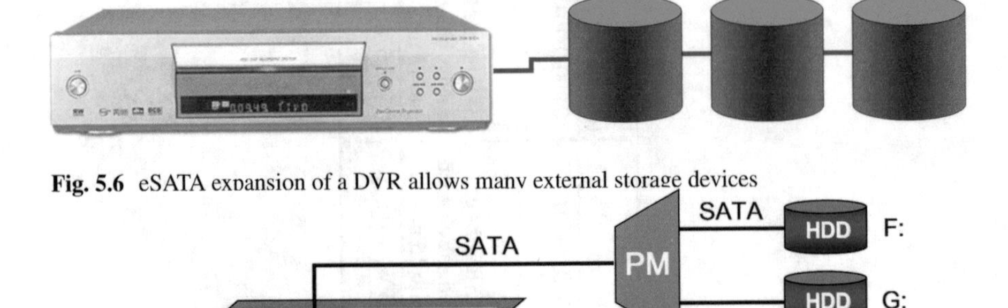

Fig. 5.6 eSATA expansion of a DVR allows many external storage devices

Fig. 5.7 Use of eSATA port multiplication to connect multiple storage devices to a single SATA host connection

port allows adding a single expansion storage device to a DVR as shown in Fig. 5.5, but a single eSATA port would allow additional expansion of a DVR storage capacity due to port multiplication as shown in Fig. 5.6. Figure 5.7 shows how port multiplication works.

Realizing that expansion of storage on DVRs is important to consumers, set-top box manufacturers such as Technicolor, Pace, and others have incorporated eSATA ports on STB offerings. This way cable and satellite companies can offer STBs with DVR capability but not a very large hard disk drive inside the DVR itself. Customers can purchase their own eSATA external storage devices that they can use to expand the capacity of their DVRs, assuming the eSATA ports are enabled.

If the eSATA chip on the STB including DVR functionality included drive locking (another feature on some eSATA electronics), then the external drives could only be used on the DVR they are attached to. Figure 5.8 shows some STBs with eSATA ports. Figure 5.9 shows a Western Digital eSATA external storage box targeted to the DVR expansion market. Other companies also offer such eSATA expansion storage devices.

Unfortunately, only a few content suppliers have made these ports active over concerns with content protection. This has dampened what could have been a hot market for expansion storage devices.

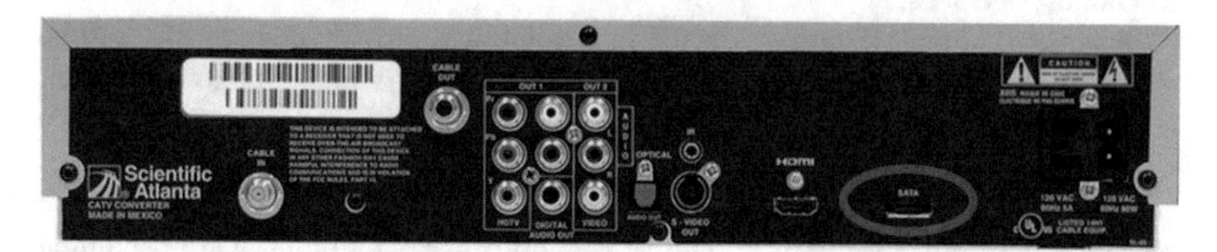

Fig. 5.8 Images of set-top boxes with DVR capability and eSATA ports for storage expansion

Fig. 5.9 Western Digital DVR Expander for adding storage capacity to DVRs

5.2.4 Network-Attached Storage for DVRs

Another option to increase the capacity of a DVR or set-top box with DVR capability is to provide a means to connect the DVR to a network. Such networking can be used for sharing recorded content with televisions, PCs, and other play-out devices or used to allow storage of video content on a storage device attached to the network, so that content can be saved even if the internal DVR storage is erased.

The TiVo DVR described at the beginning of this section provided a USB interface that could be connected to an adapter that interfaces with a home network. Some set-top box DVRs made by manufacturers have built-in Ethernet connections at the back of the box for direct connection of the DVR to a local network. Ethernet wired or wireless connections within the home can also be combined with Internet connections to allow access of a DVR to content through the Internet rather than just through on-the-air broadcast, satellite, or cable connections. Such a device combines the DVR with a device referred to as IPTV (after the Internet Protocol or IP used for Internet traffic management) that records content from the Internet for later play out. IPTV is now a popular option on Blu-ray players, set-top boxes, and TVs.

A network-attached storage device in a home can provide content that can be shared with devices in the home. Besides, just passive backup storage of content, such a network-attached storage device could be part of a home media center, which will be dealt with in a separate section of this chapter.

A variation of the network-attached storage architecture is to have a virtual drive hosted at the cable head end or in an IP network (increasingly these are becoming one and the same thing as cable moves to all IP networking). This concept is called a network DVR (or PVR), and several content companies are offering this service, sometimes as part of a video on demand or streaming content service.

Network DVR services are often promoted by online content companies to avoid the service costs of in-home set-top boxes with DVR capability. The network storage costs are covered by the cost of the content service. With the arrival of IPTV services such as Netflix, Amazon Prime, and many others; often accessed through a conventional set-top box, a TV, a game console, DVD, or Blu-ray player; many customers access video content with a combination of local storage and storage in the cloud.

In addition, some content services, such as Dish Networks, offer in-home network access (as well as mobile device access through apps) of content recorded on a set-top box on TVs not directly connected to that set-top box. Sling TV is the Dish Network technology for remote content access.

5.2.5 *Digital Video Recording Developments*

DVR products free individuals to deal with video and television on their own time rather than the timetable of the networks and other content providers. It also gives them more control of the viewing experience, for instance avoiding or fast-forwarding through ads. This makes this content more useful to the consumer. What are some other developments that could build on this useful functionality that might extend this benefit and reduce the costs and increase the availability of these devices? We will explore these potential developments here.

Consumers often would rather not erase content they have recorded to make room for new content. They also want ready access to any content that they have purchased. This gives them more control over their viewing experience since they can watch an older show whenever they wish, and they can build their own library of content. Also, the move to higher-resolution video content, for instance, the move from SD video to HD and UHD video, requires many times more storage capacity in order to hold equivalent hours of content.

A storage capacity of several terabytes (TB) to hold recorded TV content is common now with many consumers and will increase over the coming years. This could lead to the development of external storage solutions to expand the available storage of the DVR, if content owners and delivers can come to an arrangement that works for them and the consumer.

Another option is to have such huge quantities of storage located at the content provider content delivery network (CDN) and paid for by consumer service costs where they can be shared between multiple consumers.

Basic DVR capability appears to be generally in the public domain, but the use of DVR functionality with an electronic program guide is covered by patents. TiVo and Replay TV were the initial patent and IP holders for this combination of technology. DVR capability built into modern electronics usually includes some license fees in the price of the product. Building such functionality into general electronics makes integration of DVRs into Blu-ray players and other devices much easier and less expensive.

Greater integration of DVR functionality will also allow such devices to be smaller, since the circuit board can be much simpler and compact. In another chapter, we will discuss methods for integrating CE functions into the storage devices for the ultimate consolidation of electronics.

As hard disk drive storage capacity increases, 2.5 inch hard disk drives have become available with several TBs of storage capacity (as of this writing they can be as high as 5 TB in storage capacity). Using a smaller hard drive will allow smaller DVRs and devices that include DVR functionality, and in particular a 2.5 inch drive-based product could be made thinner than a DVR using a 3.5 inch disk drive and uses less power. However, 3.5 inch HDDs in DVRs are still the norm since they allow higher-capacity storage at lower cost and for static products such as set-top boxes and other DVR devices space constraints are not the primary concern for most customers, except perhaps in Asia.

As time goes on, DVR functionality has been combined with network connections and recording of streaming or downloaded content from the Internet using technology such as IPTV. As home video recording devices became capable of recording content from a very diverse set of sources, electronic program guides, and other content metadata have had to evolve as well. They are becoming more like an ongoing search agent that looks across a diverse set of sources to find the content that a customer is interested in.

Hard disk drives are the recording devices used in most digital video recorders, but some DVR type devices could be made using other storage media. There are small, often portable DVRs that use flash memory, such as flash memory cards. SanDisk introduced a product concept that they called USB TV in 2007. They worked with television manufacturers to include a special USB interface on televisions that would allow plugging in special USB-based flash memory drive that could record television content that could then be plugged into a PC to play or vice versa.

Although this device never caught on, many TVs have a USB connection, allowing them to play content from a USB device plugged into it. There are several mostly flash memory-based portable music players with AM and/or FM radio reception that allow an audio version of a DVR for radio content. Combining such functionality with Internet radio stations and podcasts as well as an appropriate content search capability could provide consumers with an enormous universe of audio content.

HDTV is now commonplace with consumers and 4 K UHDTV is ramping up. 8 K UHDTV requires even greater storage capacity per hour than HD (16X HDTV resolution is envisioned for this future standard). 8 K TVs have been on display at consumer electronic shows and should start to appear on the market around 2020 when the first 8 K video content will start to become available. Recording this content will require even larger hard disk drives both within and outside the DVR box or in large online content libraries.

In fact, as network bandwidth increases, more content will be available on demand from network DVRs or online services, reducing the need for local or external storage. The increase in resolution will drive the need for larger cache even for streamed content, but some consumers may still want to keep some content at home rather than in the cloud.

DVRs have become a standard application that can be combined with many other applications and have served as an inspiration to creating other devices for recording content and enjoying it on the customer's own schedule and using many different recording and display devices. Since resolution requirements are increasing, these devices are becoming more popular, and customers tend to keep content that they have recorded. As a consequence, the storage capacity required for these applications is increasing enormously.

5.3 Smart TVs and IP Set-Top Boxes

The proliferation of faster Internet connectivity and the development of modern data centers with large storage capacity, fast networking, and significant processing power enabled the growth of digital content delivered over the Internet. This led to services such as YouTube, Netflix, and Amazon Prime as well as Internet content available from conventional television channels. In fact, anyone with an Internet connection can serve video content to anybody who wants to watch. The trick generally is getting information on content to the people who would want to watch it.

In order to enable viewing this content, access can be made using a special IP set-top box or with Smart TVs that have this capability built into it. Figure 5.10 shows a reference design for the main board of an Android Smart TV resulting from collaboration between Google and Marvell.

Figure 5.11 shows the block diagram of this Smart TV, and Fig. 5.12 shows how the Marvell media processor SoC drives the functions of this reference TV. This design supports the features shown in Table 5.2.

Figure 5.13 shows a reference IP set-top block diagram from chip designer NXP.

An IPTV set-top box is a small computer that provides two-way communication on an IP network and can decode a video stream. IP set-top boxes have a home network interface that can be Ethernet, Wireless 802.11 g, n, ac, or some other home networking technologies. Telephone companies use IPTV (often on a ASDL or optical fiber network) as a way to compete with cable companies.

Fig. 5.10 Marvell Smart TV reference design (Google's revitalization of its android-based TV effort via Marvell SoC and reference design, Lazure, January 5, 2012, https://lazure2.wordpress.com/2012/01/05/googles-revitalization-of-its-android-based-tv-effort-via-marvell-soc-and-reference-design/)

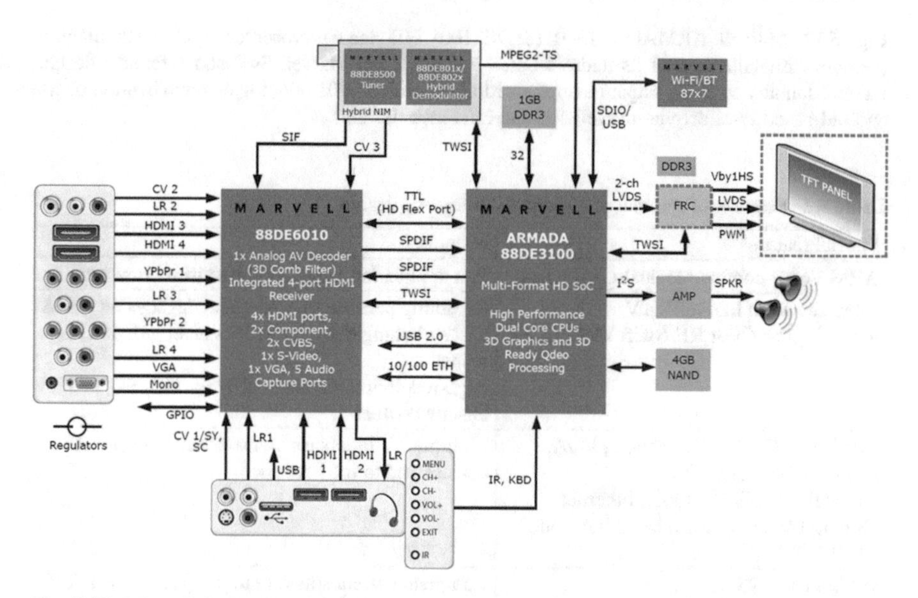

Fig. 5.11 Marvell Smart TV system block diagram (Google's revitalization of its android-based TV effort via Marvell SoC and reference design, Lazure, January 5, 2012, https://lazure2.wordpress.com/2012/01/05/googles-revitalization-of-its-android-based-tv-effort-via-marvell-soc-and-reference-design/)

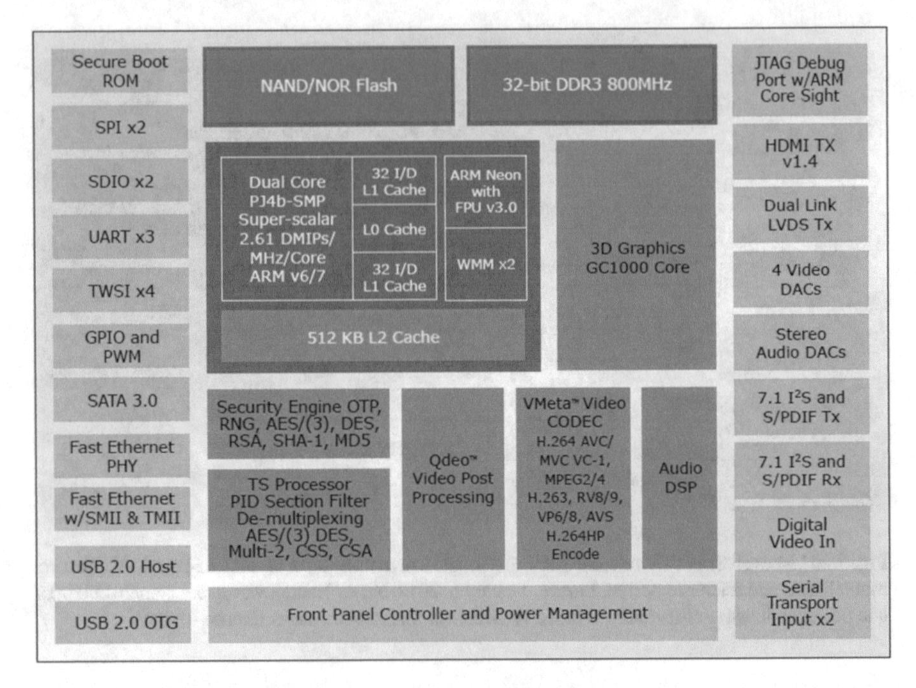

Fig. 5.12 Marvell ARMADA 1500 (88DE3100) HD media processor SoC block diagram (Google's revitalization of its android-based TV effort via Marvell SoC and reference design, Lazure, January 5, 2012, https://lazure2.wordpress.com/2012/01/05/googles-revitalization-of-its-android-based-tv-effort-via-marvell-soc-and-reference-design/)

Table 5.2 Features of the Marvell Smart TV reference design

Special features	Benefits
ARM V6/v7 compatible dual-CPU cores	Fast application launch, PC-like web browsing
Integrates with Marvell's TV technologies: Clear RF, Swift View, Qdeo	RF tuning performance matching legacy can tuners Fast switching between various analog/digital AV inputs High-resolution picture quality for HD, SD, and low bit-rate content
Inputs FR, CVBS, Component, VGA, HDMI, USB, LR audio Network: Wi-Fi/Bluetooth, Ethernet Output: LVDS, speaker, headphone out, SPDIF	Multiple AV inputs for connectivity to legacy AV devices and to the network
Android TV OS Network video on demand Bundled games Worldwide DTV middleware support Customizable reference UI	Complete system software to jump start Smart TV design and enable faster rollout of products

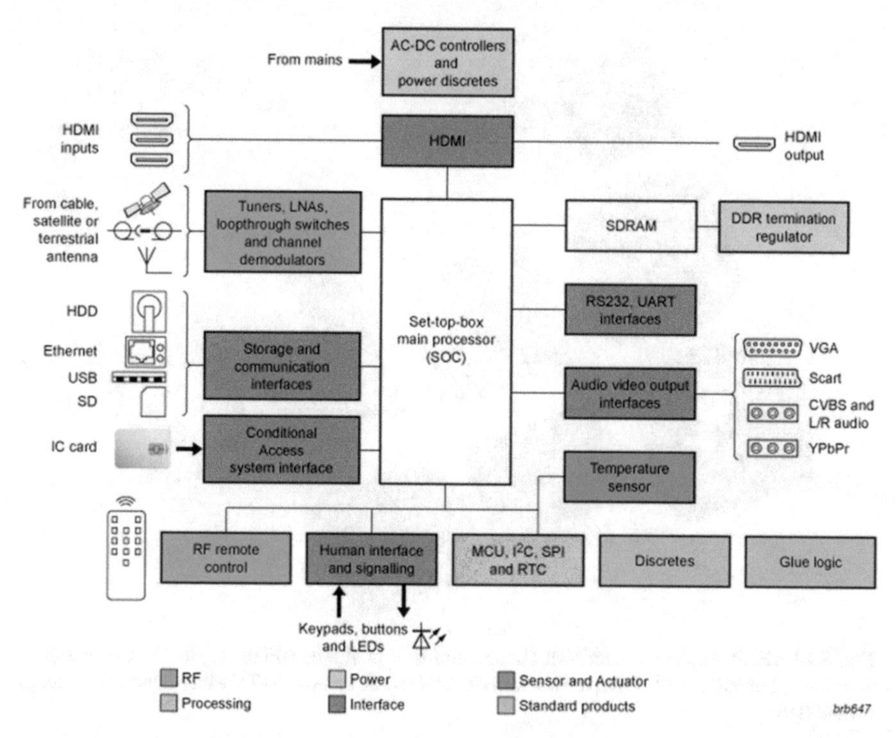

Fig. 5.13 NXP IP set-top box reference block diagram (NXP set-top box reference design block diagram, http://www.nxp.com/pages/set-top-boxes:SET-TOP-BOXES)

IP set top may receive content from official channels or allow over-the-top (OTT) services from independent providers. These services are often accessed using applications (apps) that run on the set-top box. These services include online TV channels and video on demand (VoD) and in addition to applications often provide advanced services such as HEVC, Ultra HD 4 K, multiscreen viewing, or wireless video streaming.

One example of an IP set-top box of this type is the Apple TV. Figure 5.14 shows a teardown of an Apple TV, 4th generation. Like many of these devices, it allows streaming content through the native iTunes service and downloading applications that allow viewing other content, playing games, etc. The Apple TV, 4th generation includes the Apple voice recognition application, Siri, that allows you to use your voice to control the TV.

5.4 Fixed and Mobile Game Systems

Game systems use memory and storage to store games. Home-based gaming systems that use a lot of graphics for their operation often use CD, DVD, or Blu-ray disks for game loading. Many of these systems also use hard disk drives to store

Fig. 5.14 iFixit Apple TV teardown (Image courtesy of iFixit) (iFixit Apple TV Generation 4 Teardown, September 2015, https://www.ifixit.com/Teardown/Apple+TV+4th+Generation+Teardown/49046)

game graphical video content so that it can be accessed quickly. The hard disk drives used for these applications are increasing in storage capacity but are usually relatively low capacity so they can be as inexpensive as possible in order to meet the bill of materials cost requirements.

Multiplayer games done through the Internet could change some of these storage requirements, particularly for the home-based game systems. This could lead to a greater amount of storage on a network and especially on the Internet that gamers access. However, although many lower resolution games are just run over the Internet, higher-resolution games, even if involving connected players, require good amounts of local storage capacity.

5.5 Home Media Center and Home Network Storage

A *media center* is a computer adapted for recording or playing content such as music, videos, and photographs out of a local hard disk drive or other storage devices or over a network that is connected to some storage device where the content is resident. In the mid-2000s, companies such as Microsoft and Apple supported media servers in their operating systems and many computer companies, such as HP-created computers with media center capability. In addition, the open-source XBMC media center software was gaining popularity. By the late 2000s, the major operating systems (Microsoft and Apple) had dropped their media server support.

The reason that media center PCs became less popular is the rise of multi-platform applications, particularly on mobile IOS Android operating systems and TVs running Amazon's Fire TV, Google's Chromecast, Smart TVs, and Roku. In addition, home gaming consoles like the Playstation and Xbox enable media serving applications. This has enabled content to be more accessible for mainstream consumers, while media centers are used by technology enthusiasts to manage their own content using Plex or Kodi (the new name for XBMC). However, media centers are still in use and have played an important role in consumer electronics.

Media centers are often also able to play content from CDs, DVDs, Blu-ray disks, or other distribution media. Like many other entertainment devices in the home, they usually have a remote control that can control the function of the device, and the content can be viewed on a TV set.

The media center can be a device designed for this function or a regular computer (such as a PC or game console) with appropriate software (also there are small inexpensive computers such as the Raspberry Pi that can be used for media centers). Media centers usually have a DVR or PVR function built into them to record video on hard disk drives. Media centers are also a natural venue for IPTV where video and other content is taken off the web. Media centers can also be a home gateway for Internet of Things devices in the home, used for automation and entertainment.

Media centers can be created as single appliances providing network connectivity as well as content storage or created using separate network storage devices and digital media adapter products. Dedicated digital media adapter products have been available from many networking and computer hardware companies. For the sake of simplicity in this section, we will refer here to single media center appliances with separate storage and media adapter combinations as well as multiple use devices with storage (such as personal computers) that can act as media adapters and media centers.

Some of the software that can be used to transform a PC into a media center is given below:

- Kodi
- Movavi
- MediaPortal
- Plex
- Emby
- MythTV

5.5.1 *Basic Layout of Media Center Devices*

Various commercial devices have been available for home and commercial markets offering media center capability. Figure 5.15 shows some historical devices from the mid-2000s from HP, Samsung, and Cannon. Higher-capacity products have been available from several other vendors including AMX (now part of Harmon/Samsung). Some storage companies, such as Western Digital and Seagate, offer devices for storing and streaming content.

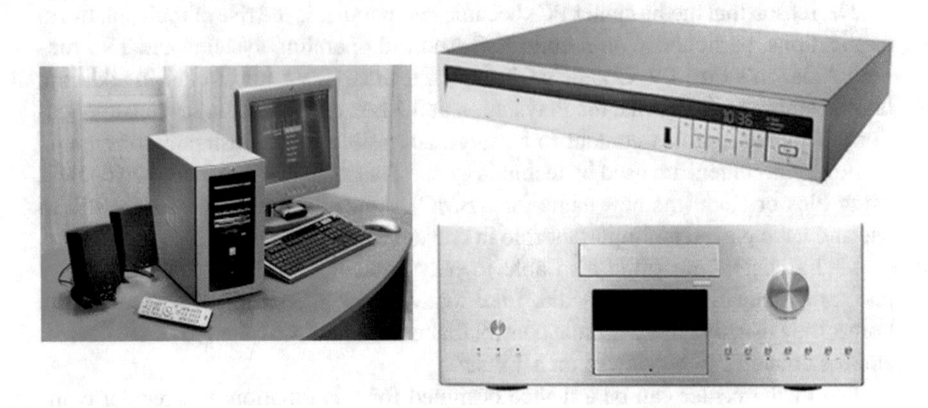

Fig. 5.15 Media centers (http://www.tvsnob.com/images/moxi_samsung_media_center.jpg, http://www.broadbandhomecentral.com/report/backissues/images/HP_MEDIA_CENTER_PCb.jpg, http://ww1.prweb.com/prfiles/2005/09/14/285332/LXMediaCenterLarge.jpg)

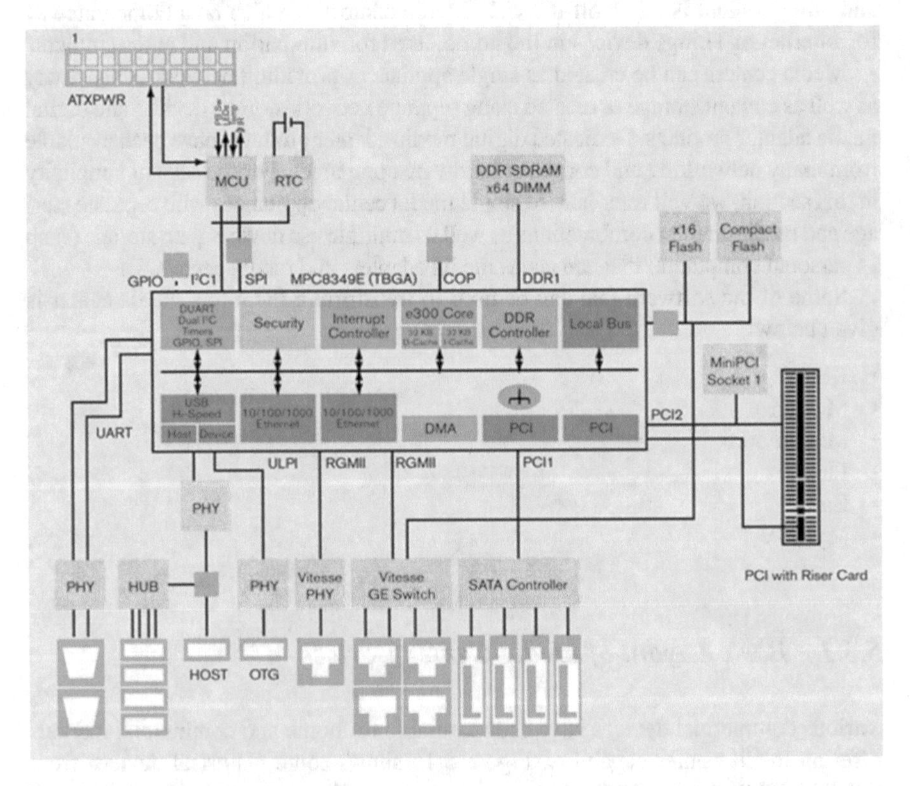

Fig. 5.16 NXP media server-in-a-box reference platform block diagram (http://www.nxp.com/products/microcontrollers-and-processors/power-architecture-processors/powerquicc-processors/powerquicc-ii-pro-83xx/mpc8349e-mitx-media-server-in-a-box-reference-platform:RDMPC8349EMSIB)

Fig. 5.17 NXP media server-in-a-box reference platform

Figure 5.16 shows a block diagram of an NXP media server reference platform. Figure 5.17 shows the actual reference platform electronics. A media center must provide the performance necessary for the deployment of functions like MPEG2/4 decode and encode, digital video recorder (DVR) combined with an electronic programming guide, gaming, Internet browsing, and digital content editing. In addition, the media center may also act as a media server to clients throughout the home, as well as provide voice and video conferencing. In order to be deployed in a home environment, the system design must be streamlined to blend into the home environment.

Designers of media centers must integrate a number of connectivity technologies, high-performance processing capabilities (using embedded microprocessors), and a number of digital media (video and audio) processing technologies.

Some basic design features can significantly affect media centers and their general usefulness. It is important to keep these features and options in mind when designing these products:

1. Will the media center try to connect various devices in the home, including computers and consumer electronic products?
2. Will the media center connect to other networks in the home such as coax cable, phone lines, or power lines?
3. How well is the device able to handle commercial as well as personal content?

4. Will the storage of content be inside the media center or on the same network as the media center?
5. Can the media center play content located in other devices and if so which other devices can share this content?
6. How will commercial and personal content be protected and how can privacy be ensured?

We will deal with these questions in the following sections.

5.5.2 Home Networking Requirements for Media Centers

A media center is basically a computer with electronics and software designed to support multiple streams of content in a home environment. It can look like a personal computer or some sort of custom-built device designed for the home environment. The digital storage requirements depend upon the amount of content that the system will contain. Since this could include both personal and commercial content, the actual storage required could be very large.

In addition, media centers can act as a DVR. Thus, the storage requirements for DVRs described earlier may be true for home media centers with additional storage capacity required for personal content.

5.5.3 Home Media Centers and the Internet

It is natural that a home media center network should also connect to the Internet. Connecting to the Internet allows the home media center to access commercial content through services such as IPTV and also personal content that other people post on the web to share. Content sharing on the Internet is one of the fastest-growing applications and one that consumes quite a bit of bandwidth and storage in the data centers that handle content sharing. Security of commercial content and protection of personal content are important for the continued growth of this market. A universal media center or home gateway must be designed to access and use content whatever the source is.

IPTV describes a technology allowing a home entertainment system to access video and other contents through the Internet, downloading and viewing content when it is desired. IPTV has time-shifting approaches similar to DVRs and can result in significant digital storage requirements on the media center as content stored builds up. As discussed earlier IPTV can result in external storage demand for longer-term retention of recorded content, or this content may remain in the cloud to be accessed when needed.

An IPTV set-top box or a home media center incorporating IPTV may also support other functions enabled by Internet connectivity. An important feature for an

Internet connection is that data can be transported both directions, allowing for built-in interactivity options that are not typical in cable or satellite set-top boxes. This interactivity can be used to make an IPTV device capable of other Internet functions such as voice over IP.

IPTV and other Internet content access methods are very popular and have become larger than traditional content distribution technologies. Sharing of personal content may also include a centralized home repository for content such as a home media center, or it may use content stored in the cloud. As time goes on all these, connections and networks could merge in some fashion, combining these features in a seamless service offering. That is, if it can be done simply and is easy to use—this may be the biggest challenge and could take some time to perfect.

5.5.4 *Future Media Center Capability*

The major factors driving media center capability over the next few years are the following:

- Higher-resolution and dynamic range content requiring greater storage capacity and internal bandwidth requirements
- DRM and privacy protection of personal content
- Indexing and organizing content to make it easier to find and use material
- Unified backup of content on a media center including deduplication of similar files
- Use of the Internet to back up personal content stored on a home media center to prevent its destruction in case a disaster destroys the media center or home gateway

We will explore these factors below, which are important for any content device, including home media centers.

5.5.5 *Faster Organization and Content Search in Home Media Centers*

We will have potentially vast amounts of stored content in our home environments (both commercial and personal content in a home or in associated cloud storage could reach petabyte levels before 2030), we will increasingly need some way to find this content. There are three keys to the efficient access of potentially vast stores of home content:

- Efficient metadata creation to describe the content to a search engine
- Effective indexing and context placing of content to speed up searches
- A way to display the results of a search so the user finds the content closest to what he/she was searching for

Let us look at these three requirements one at a time. *Metadata* is information about a data file that is stored with the data file and can be accessed by external programs for various uses. Metadata can be entered manually such as with editing of content where a description of a picture or a scene is added to describe it. This is still the most metadata some consumers have on their content (especially older content), and because of the effort required with manual entry, the creation of most metadata today is rather primitive.

Automatic metadata generation, where the information about the data file is derived from the content in the file, will make the use of metadata easier. Such technologies are now available, and they are revolutionizing the usefulness of new and historical content. Image recognition technology using artificial intelligence algorithms is enabling increasing sophistication in automated metadata generation. With such powerful metadata creation tools, a user is able to create metadata on all of this historical content as well as new content, making the information available for rapid access of content considerably greater.

For text this could consist of keywords or phrases that give the gist of the material, and for audio this could be keywords or phrases derived from a text-converted version of audio as well as useful descriptions of nonhuman sounds. For still or moving images, metadata includes image elements that allow recognition of people, animals, or places and for the moving image some sort of synopsis of what action occurs, involving which elements. For all these data files, date and location information (e.g., if available from GPS) are also useful information.

Indexing of content uses metadata and keywords about individual files helps speed up searching of this content. As we extend this capability to correlate multiple images and videos and identify individual visual and audio elements, these tools could become even more useful (note, for instance, that Facebook as well as many mobile devices can recognize people in photographs). Much investment and research is going into developing such capabilities. Benefits to users are enormous.

5.5.6 *The Future of Home Media Content Access*

As a centralized repository and access point for a home's entire user-generated as well as commercial content (likely combined with cloud-based services for heavy processing and correlation and an online library), a future home media center or home gateway must:

- Be easy to use.
- Be able to provide all required storage and sharing services.
- Have the bandwidth and storage capacity to meet household needs, and additions of storage and bandwidth must be easy to install.
- Have software capable of providing proper control of content access, security, and privacy of content and provide automated protection of the content for simple file corruption to complete home disasters where the home media center and its local backup itself may be destroyed.

- Coordinate with cloud-based content repositories and services.
- Be inexpensive enough in high-unit volume shipments, that it can achieve a mass market.

Even if the general architecture that is called a home media center does not become the standard way that content is handled in the home, these requirements still hold for consumer content.

In order to incorporate these capabilities, various technologies will likely be integrated into the centralized home storage repository and control center:

- DRM or other content security as appropriate
- Means to protect personal content privacy such as storage device encryption (as proposed by the Trusted Computing Group)
- Networking capability providing adequate bandwidth to support many streams of ever higher-resolution content in the home (including wireless networking)
- Storage devices providing the proper storage capacity at the point of use and access to the centralized content so this data can be shared refreshed and controlled
- Automatic metadata creation of text, audio, and video content including context placement capability to enable rapid search
- Online services offering various services including backup, protection, sharing, and rapid recovery of home content

5.5.7 Backing up and Disaster Recovery for Home Media Centers

Home media centers or other home content devices will contain some of the most valuable information in a home. A centralized repository for family digital photographs, videos, and other personal contents as well as DVR recordings and purchase and downloaded content from the Internet and other sources will have a great value to home users. The loss of some or all of this information can be a nuisance at least, and in the case of user-generated content, the irrecoverable loss of content would be tragic—like losing the family photograph album in a fire!

For this reason, it makes sense that a home media center, as a likely aggregator for valuable personal and commercial content, should have a means of backing up the data it contains as well as protecting this content from a disaster that may destroy the digital storage devices where this content is stored in the home.

Backup of a home media center should be an automatic function of the device, once set up it proceeds without frequent intervention by the user. The setup itself should be easy so even a nontechnical user can quickly implement the most common basic data protection with additional capability for more experienced users to do more if they wish (although this has its own dangers since if nontechnical users are forced to use this function and it is not well designed, they could get lost and even do damage).

Ideally as soon as a device is connected to the network where a home media center exists, it should back up that data and be able to restore it to the original device (or its replacement) when needed. The home media center should also make it easy to move content from one device to another such as when hardware is upgraded or changed. This content should be secured from outside intruders through encryption and access control.

A further step in the protection of user content (both personal and commercial) is to create an online service where the content can be automatically backed up to a secure location on the Internet. This location must be secure and prevent unauthorized access to the content. It must be set up to do incremental backups of the content on a regular basis, and if needed an individual file or even entire household data recovery should be easy to perform. Today, there are many storage services available that enable online backup.

5.5.8 High-Resolution Content for the Home

Table 5.3 compares data rate and required storage capacity per hour of content for SD, HD, 4 K UHD, and possible 8 K UHD resolutions for broadcast and cable content. Note that actual numbers may vary depending upon the compression used. Compression has gotten more sophisticated as higher power processing costs have declined, and techniques such as HVEC that require more encoding processing allow considerably higher compression than older format. Internet data streams are usually compressed even further to better utilize available bandwidth.

On the other hand, the bandwidth for a stream of a given resolution from an optical disk or other physical distribution formats may be twice as those shown in this table. Although higher compression ratios are possible today with modern formats, as content increases from SD to higher resolution, the bandwidth and storage capacities increase.

A media center must be able to handle multiple streams of content in order to serve its function of sharing content and also must support *trick modes* such as fast-forward and reserve. These trick modes increase the required bandwidth of the system by a factor of 3 or more if continuous streaming must be supported at the same time. In practice media centers are often designed with the equivalent of 16 separate channels of content. Table 5.4 shows the resulting bandwidth required for 16 channels for the same three resolutions. The numbers are approximations since compression rates and resolutions may vary somewhat.

It is clear that the 8 K Ultra HDTV media center with 16 streams would have a 45 MB/s total bandwidth requirement. With other network traffics this could tax one-gigabit Ethernet networks and require a higher bandwidth wireless or ten-gigabit wired Ethernet network. Wireless networks with this bandwidth are becoming available and should be more common in the future.

Table 5.3 Video resolution, single-stream rate, and storage capacity required per hour for a stream

Video resolution	Stream rate (Mbps)	Storage capacity/hour (GB)
SDTV	~2.0	~0.9
HDTV	~5.0	~2.3
4 K UHDTV HEVC	~15	~6.7
8 K UHDTV HEVC	~25	~11.3

Table 5.4 Bandwidth requirements for a 16 channel media center as a function of video resolution

Video resolution	16-stream rate (Mbps)
SDTV	~32
HDTV	~80
4 K UHDTV HEVC	~240
8 K UHDTV HEVC	~400

5.6 Chapter Summary

- Digital video recorders (DVRs) have become a common home consumer device, often incorporated into set-top boxes (STBs). DVR expansion and content sharing have led to efforts to create external direct-attached and network-attached storage systems to provide additional storage capacity to these devices as well as allowing content sharing.
- eSATA is used for storage expansion of DVRs since the current standard supports up to 6 Gbps data rates and daisy chaining of eSATA storage devices allowing additional storage expansion.
- Many different options for home media networking to support sharing of DVR content are likely, including coax cable-based systems, wired and wireless Ethernet systems (such as 802.11x wireless networking), and possibly power or phone line-based networking systems). The choice will depend upon the devices being connected, what is supported by one's service providers and local connectivity options.
- IPTV services are displacing many traditional content delivery networks. In addition, multistreaming services and potentially network PVRs may reduce the amount of commercial content actually stored in the home. Hobbyists are still interested in home media center services that allow them to manage and store their own content libraries.
- A key reason for a home network is backup and protection of data and content in the home. In addition to in-house backup, there are many Internet-based service providers that can back up or synchronize the data on computers and other devices in the home. Such remote backups can provide a path to disaster recovery of personal, business, and commercial content in case of an accident that damages or destroys the products where the original content was stored.

Chapter 6
Storage in Automotive and Mobile Consumer Electronic Devices

6.1 Objectives in this Chapter

- Look at the functions, design trade-offs, and use of digital storage in common mobile consumer electronic products.
- Examine the unique and demanding environment of digital storage for entertainment and navigation in automobiles and how to design systems made for these environments.
- Learn what sort of storage capacities will be required for mobile players in the future and how to choose the best type of storage device from the mobile storage hierarchy.
- Find out how digital storage is integrated into still and video digital cameras, AV players, mobile phones, and other mobile consumer products.
- Discuss potential future applications of storage in mobile devices, including life logs.

This chapter extends the analysis of the prior chapter into the design and digital storage requirements of consumer devices in the home to mobile consumer devices. Applications used outside the home such as in automobiles, mobile music and video players, and digital still and video cameras have significant challenges with environmental reliability, battery life in normal use, and resolution requirements. These characteristics must be taken into account to design the right type of digital storage and memory for these applications.

This chapter explores the unique drivers in mobile consumer product digital storage both at the present time and likely in the future. It gives some insights into what sort of storage capacity will be available and how much this will cost for the available storage options. It ends by briefly considering new uses of digital storage in mobile devices that could increase the required storage considerably both in these devices and in the home.

T.M. Coughlin, *Digital Storage in Consumer Electronics*,
https://doi.org/10.1007/978-3-319-69907-3_6

6.2 Automotive Consumer Electronics Storage

Automobiles represent one of the most challenging environments in consumer electronics. I was raised in the upper Midwestern United States, and I experienced temperatures that ranged from as low as −30 °F (−34 C, without including wind chill) to as high as 125 °F (46 C). In a car, temperatures can get even higher, with the windows rolled up. Furthermore, electronic products designed for this environment must also deal with the regular vibrations, accelerations, and shocks that are common to automobiles.

Storage devices designed for these applications must be very robust to survive in such an environment. Because of the extended life expected for embedded automobile components in such a harsh environment, qualification times tend to be rather long, often better measured in years rather than months. Also, products once qualified must be supported and components available for servicing for 10 years or more.

This section will explore the environmental requirements for automobile applications as well as storage devices designed to be used in automobiles. Common storage devices used in automotive consumer applications are optical disks, hard disk drives, and flash memory.

6.2.1 Digital Storage for the Automobile

Table 6.1 shows some characteristics that could be expected for storage devices in an automobile environment. These are not comprehensive, and many of these environmental conditions such as shock and vibration are reduced at the location of the electronic device to increase the reliability.

Automobiles are incorporating more and more electronics. Many automobiles run at least partially off of electricity rather than gasoline, and in the future electric and hybrid electric gasoline or hydrogen automobiles will be in the majority. Electronics is involved in automobile control systems, safety and diagnostic systems, as well as navigation and entertainment systems. These trends are accelerating with the appearance of automated driver assistance systems (ADAS) and increasing levels of autonomous driving.

Most digital storage in automobile control, diagnostic, and safety systems are embedded solid-state memory, since these storage capacity requirements are generally not too great and the robustness required is high. Automobile navigation and entertainment systems are the exception because they could require very large amounts of storage. This storage could be removable or fixed depending upon the purpose it is used for.

Navigation systems use digital storage devices for two basic purposes:

- To supply data to the navigation system on basic routes, locations, and maps

Table 6.1 Automobile environmental conditions of interest for most CE applications

Environmental factor	Range
Temperature	−40–150 F (internal automobile temperature)
Relative humidity	10–100%
Vibration	Can be very severe, usually damped around electronics
Shock	Can be very severe, usually absorbed around electronics

Table 6.2 Comparison of digital storage requirements for various resolution maps

Description of map	Storage capacity needed (MB)	Source
California highway map (PDF)	7.73	http://www.metrotown.info/
Santa Cruz County, California, topographic base map	9.8	http://www.usgs.gov/
Salinas, California area Landsat image	~55	http://www.nasa.gov/

- To acquire and store real-time traffic updates and suggest alternative routes when appropriate

The first approach uses relatively fixed data that changes slowly with time. Data for these applications was often supplied on a removable media such as an optical disk (a CD, DVD, or Blu-ray), but with newer connected cars, these updates can happen over a consumer's Wi-Fi network or cell phone network. This data is downloaded into a mass storage device in the automobile that may use a hard disk drive or a flash memory device. This map data is evolving from simple 2D maps to what are called 3D maps that can contain much more information about the local environment, including actual photographic route images and topographical features. 3D maps require a considerably higher-capacity storage.

If audio or video clips related to possible destinations are included, then the storage capacity for even this relatively static content can become rather large. It is easy to see that a simple static navigation system could grow to be many gigabytes in size. Navigation is often incorporated into the car's audio entertainment system. This combination is called an automotive infotainment system.

As the storage capacity of optical storage (or the internal mass storage of the infotainment system) grows, automobile navigation will become much richer. Table 6.2 gives some idea of the amount of storage required to store various resolution maps of a geographic region.

Real-time traffic update systems for automobile navigation first become popular in Japan due to the narrowness of the streets in Tokyo and the propensity for traffic to become congested. For this reason, *Vehicle Information and Communications Systems* (VICS) became very popular there. Several million of these systems have been installed in automobiles in Japan and other Asian countries. These in-car navigation systems are combined with a sensor and radio network located on the streets of the town that give real-time traffic condition information and, with the help of the navigation system, provide the user with alternative routes to avoid congestion and delays. The real-time route updates are stored on an on-board storage device.

With the rise of various navigation mapping applications, running on mobile phones or on an inboard infotainment system, real-time traffic data tied to machine algorithms running in the cloud that provide drivers the best route to get where they are going are common in urban areas in many cities around the world.

Usually the storage devices used in on-board automotive infotainment systems are ruggedized hard disk drives that also contain map, route, and perhaps destination information. These hard drives allow for rapid updating of the route information as well as ongoing changes in the general map and destination information. It is likely that hard disk drives could be augmented by flash memory write cache (a hybrid drive) or even substituted by a solid-state drive in future navigation systems, depending upon the storage requirements and the price of flash memory.

6.2.2 Basic Layout of an Automobile Infotainment System

Let's look at some examples of schematics for an automobile entertainment and navigation (infotainment) system. Figure 6.1 shows an automotive infotainment system block diagram. The infotainment system combines entertainment, multimedia, and driver information functions in one module. It offers AM/FM or satellite radio, CD/DVD/Blu-ray player for music and video, navigation system, data and multimedia ports (both wired and wireless), as well as vehicle status.

The power supply is connected to the 12 V or 24 V source power and changes the voltages to meet the needs of the DSP, memory, and other ICs in the module. In some cases, there may be ten different power rails. The high-speed CAN (up to 1 Mb/s) is a two-wire fault-tolerant bus commonly found in automobiles. The audio DSP performs I/Q demodulation and outputs digital audio and data.

Figure 6.2 shows a teardown of a BMW i3 digital radio module. In this teardown, a special 2.5-inch hard disk drive designed for automobile application is used for the mass storage of navigation and entertainment content. Figure 6.3 shows the schematic diagram for the main board in the BMW i3 digital radio module, and Fig. 6.4 is the 200 GB hard disk drive used as the mass storage in this device.

This particular hard drive came from Toshiba. Capacities for these drives are up to 320 GB. This drive has a rotation rate of 4200 RPM with an internal transfer rate up to 976 Mb/s and an average seek time of 12 ms. It can handle operating temperatures between −22° and +185° F and nonoperating temperatures between −40° and +203° F. The drive can operate at altitudes up to 18,536 feet and withstand up to 3G (29.4 m/s2) vibration during operation.

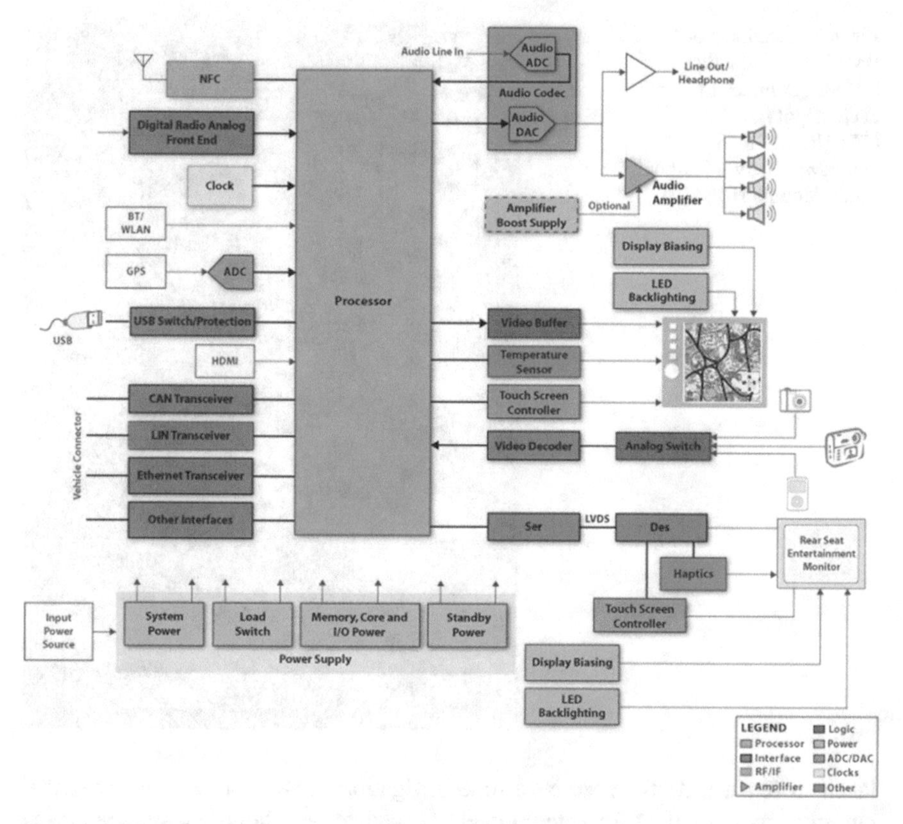

Fig. 6.1 Automotive infotainment block diagram (Courtesy Texas Instruments)

6.2.3 *Storage Device Trade-Offs and Options for the Automobile*

Automobile entertainment systems have been transitioning from optical disk formats such as CDs, DVD, and Blu-ray disks[1] to store music and movies to on-board mass storage, with content moved to the automobile using wireless networks. Also, as mentioned earlier, destination audio and video information could be available through the automobile infotainment system to learn more about a destination before reaching it.

A real-time navigation system requires a digital storage media that can be written and read many times. These functions are popular in places where sensors in the streets detect, or real-time human input reports, local traffic and road conditions. This information is used to tell subscribing automobiles what route should best be

[1] The use of consumer DVDs and Blu-ray disks in automobiles required the addition of the DVD Copy Control Association's Content Scramble System into the Automotive DVD Playback System.

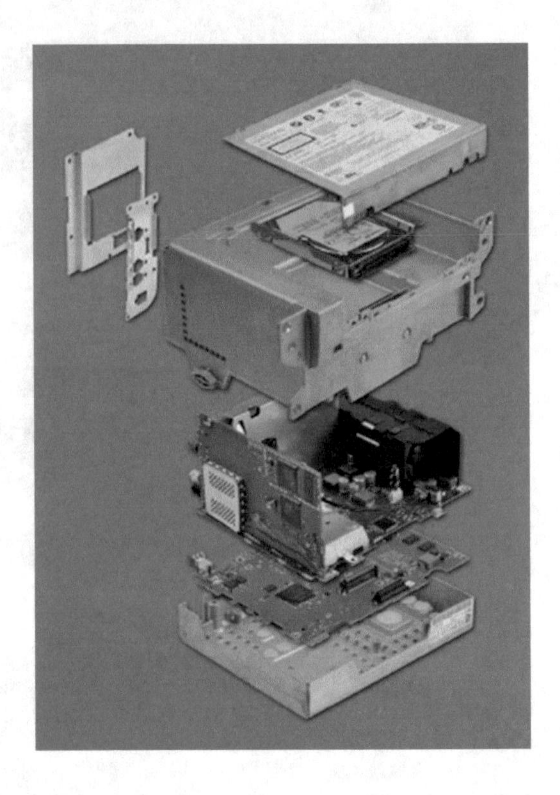

Fig. 6.2 Teardown of a BMW i3 digital radio module (Courtesy of TechInsights) (TECHINSIGHTS Teardown, BMW i3 Digital Radio Module HBB125)

taken to reach a destination. Real-time navigation systems as well as ones that requires lots of content for entertainment or navigation often use a storage device such as a hard disk drive (HDD) but may also use a solid-state drive (SSD).

HDDs used in automobiles live in a very harsh environment, very different from those in computers and even most consumer applications as discussed with Fig. 6.4. In order to operate under such temperature extremes and the vibration and shock conditions found in automobiles, the track density and linear density of HDDs are reduced to give lower storage capacity but a more rugged storage product with performance margin to spare. For very extreme temperatures where HDDs (or SSDs) might be expected to operate, they must be heated or cooled to stay within specified environmental tolerance.

A shock absorbing case can be used around a hard disk drive to reduce shock and vibration sensitivity. There have been several designs of such cases. In many infotainment systems, the shock absorption is built into the infotainment module.

Optical disks have been a popular storage media for automobile applications, but these applications are transitioning to using hard disk drives or flash memory. Although the flash memory device is more robust at extremes of temperature and shock, the price per GB is three to five times higher and likely to remain so for several years to come.

Although the various power specifications for the hard disk drives appear to be higher, in actual use the hard drive stays on less than 5% of the time (low duty cycle),

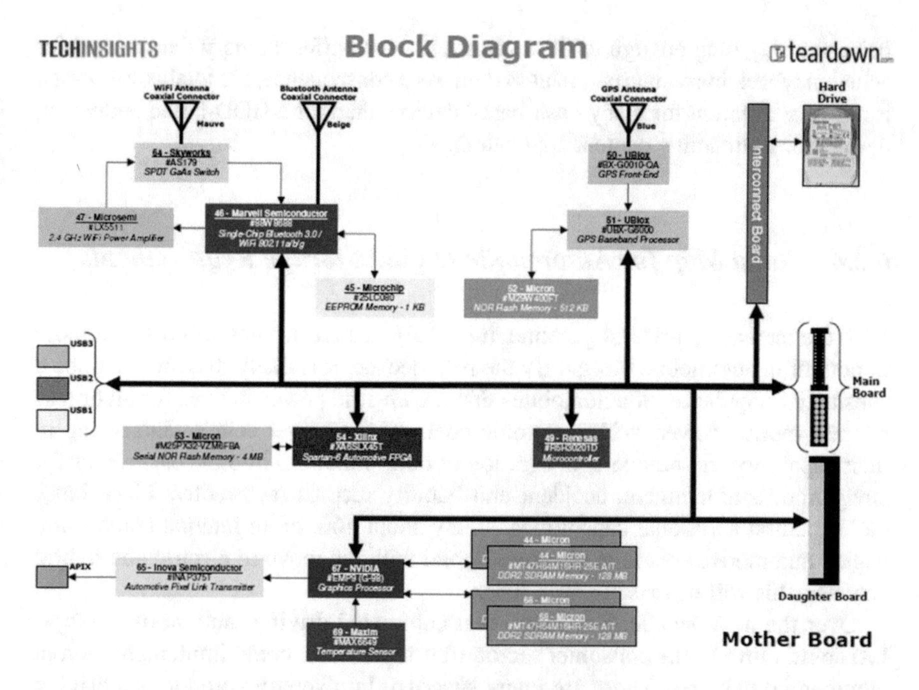

Fig. 6.3 Partial schematic diagram of the BMW i3 digital radio module (Courtesy of TechInsights) (TECHINSIGHTS Teardown, BMW i3 Digital Radio Module HBB125)

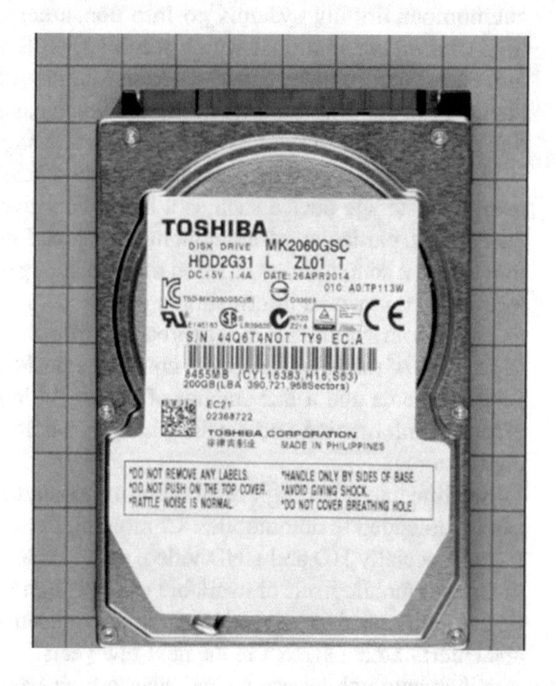

Fig. 6.4 Hard disk drive used in the BMW i3 digital radio module (Courtesy of TechInsights) (TECHINSIGHTS Teardown, BMW i3 Digital Radio Module HBB125)

being on only long enough to fill a semiconductor buffer memory that is used for actual playback in an entertainment system. As a consequence, the total power usage is not very different for many flash-based devices than for a HDD-based equivalent device, for a streaming content application.

6.2.4 Road Map for Automobile Digital Storage Requirements

With the increasing price of gasoline, it is likely that electronics will become more important in automobile design. By the next decade, it is likely that there will be a substantial population of automobiles that use electric power sources for all or part of their motive power. With electronic controls and power sources increasing in their capability and numbers, integration of other functions into the electronics for navigation, entertainment, accident and liability records (automotive black box), etc. is natural additional capabilities. Today about 40% of an internal combustion engine automotive cost is in electronics, and with the move to electric and hybrid vehicles, this will increase to over 70%.

Over the next decade, it is likely that automated driving (automotive autopilot) could move to the consumer sector after more widespread implementation in commercial vehicles. There are many successful university autonomous driving programs as well as major industry programs at companies such as Google, Uber, and Apple, as well as new and traditional automotive companies. Once such autonomous driving systems go into consumer transportation, riders will have time to do more than just watch the road while they drive. This opens up even more need for communication and entertainment technology to use this time productively or for entertainment. These changes and the increasing amount of time that people spend in their cars will drive new requirements for digital storage.

Real-time traffic and road updates using a wireless network or radio system and a rewritable storage device such as a hard disk drive or flash memory are becoming widespread, particularly for autonomous vehicles or autopilots. As an aid in navigation and as a source of information to the passengers, it is likely that information on locations, commercial information on the area, and other information, automatically recorded in an automobile as it approached a destination, will become more common. Base maps of roads will probably consume considerably less than 1 GB and cover a very large area and a fair amount of detail. As indicated in the table earlier, topographical information and satellite images would require much larger amounts of digital storage.

Satellite and other high-resolution images and even video taken from a location can be uploaded to automobiles. Combining these needs with traditional entertainment, especially HD and UHD video, could require fairly significant storage either in the automobile itself or available to it through wireless networking. It is not difficult to see a need for several hundred GB or even a TB in an automobile navigation and entertainment system in the next few years.

When automobiles are parked near to their base (your home), they need to connect to your home network, probably using a Wi-Fi network interface or possibly a cell phone network. This will allow updating and synchronization of content between your home and your automobile. The automobile will likely only contain copies of a subset of your entire data since the larger bulk of your content is available relatively conveniently either from your home network or the cloud.

If DRM issues can be resolved, it would be very convenient for consumers to be able to copy their CDs, DVDs, and Blu-ray disks to their home network and make them available in their automobiles, without having to handle physical disks that can become damaged and lost. A much larger library of content can be used and organized for easier use when they can be put on a network storage device or cloud storage. Thus, in the long run, I expect that optical media will not be the preferred way to use content in automobiles.

Hard disk drives and flash memory are more likely to be the primary storage in automobiles in the future. Flash is attractive because it has a greater operating temperature range, and it is more shock and vibration insensitive. However, it appears likely that flash memory will remain about three to five times more expensive than an equivalent storage capacity available on a hard disk drive. For large content storage, even in an automobile environment, there is a role for hard disk drives for some time to come, particularly if content sizes continue to increase.

Flash memory may be a way to transfer content in and out of a car (like a USB drive) or be used for certain applications within the automobile that need a more robust storage product. I also consider it likely that hybrid storage devices that contain a flash memory cache built into a hard drive or on the car electronics motherboard could provide some of the faster response under extreme environments that flash can provide while adding only a little to the cost of the hard disk drive.

6.3 Mobile Consumer Devices

MP3 and video portable players were some of the most popular stand-alone consumer devices. These products stored local copies of music, photographs or video content and so digital storage is an important element in their design. Now these functions tend to be built into our smartphones or tables, and stand-alone product sales have plummeted.

In this section, we will describe the trade-offs in design and performance that must be made in battery powered consumer devices. Many of the observations that are made here will also be applicable with other mobile devices.

6.3.1 Mobile Consumer Electronic Designs

We will investigate three types of portable consumer electronic devices and the digital storage and other design requirements for these devices. These are (1) mobile media players, (2) smartphones, and (3) electronic tablets. All of these devices can store personal as well as commercial content.

6.3.1.1 Mobile Media Player

A media player is a simple device in concept. It consists of a sound system that is often an amplifier with a headphone jack, connected to a codec that decodes data read back from a storage device (it may also store video as well as audio content and have a display for this purpose).

Typically, a battery provides the power for the device. Since the power from a battery is limited, power consumption is a critical constraint in the amount of time that a customer can use a device between charges. Thus, power saving modes and various other techniques are used to reduce power consumption and increase the time that the device can be used on a single charge. In addition to the power requirements, customers often prefer to have devices that are more compact, have a lighter weight, are more durable under likely usage conditions, have an easy-to-use way to obtain content in the device, look good, and of course are within the budget of the user.

Figure 6.5 shows a block diagram of a mobile music player (with voice recorder capability).

Most mobile portable media players are "temporary" storage devices that contain a copy of content that is copied from some other device owned by the consumer such as a computer. Since the player can be lost or damaged, it is wise to retain the "original" copy on a static device that stays in the home of the user where it is hopefully less subject to damage and loss.

The MP3/voice recorder player performs noise reduction, speech compression, and MP3 stereo recording/playback at all rates and options required by the MPEG/speech compression standards. The following is a description of the components for such a device.

- The *digital signal processor (DSP)* performs the audio encode functions and executes post-processing algorithms like equalization and bass management and system-related tasks like file management and the user interface control.
- The *memory* stores executing code plus data and parameters such as content metadata.
- The *peripheral interface* allows the user to control input/output (I/Os) and the display.
- The audio *encoder/decoder* or *codec* interfaces with microphone, radio signals (if the device included a radio receiver), and other audio input and with the headphone and possible speaker for audio output for digitizing the audio in the DSP.

Power conversion changes the provided power to serve run various functions in the device.

Figure 6.6 is an image of a teardown of a portable music player.[2]

[2] iFixit 2013 Teardown—ifixit.com.

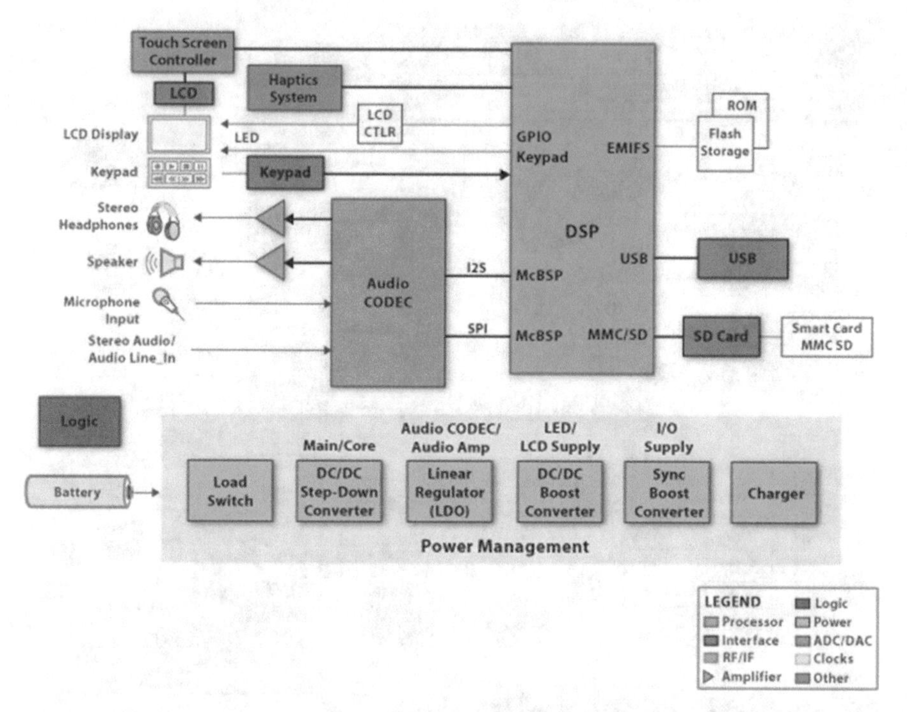

Fig. 6.5 MP3 player and voice recorder block diagram (Courtesy of Texas Instruments) (Texas Instruments, MP3 Player/Recorder, TI Reference Design Library)

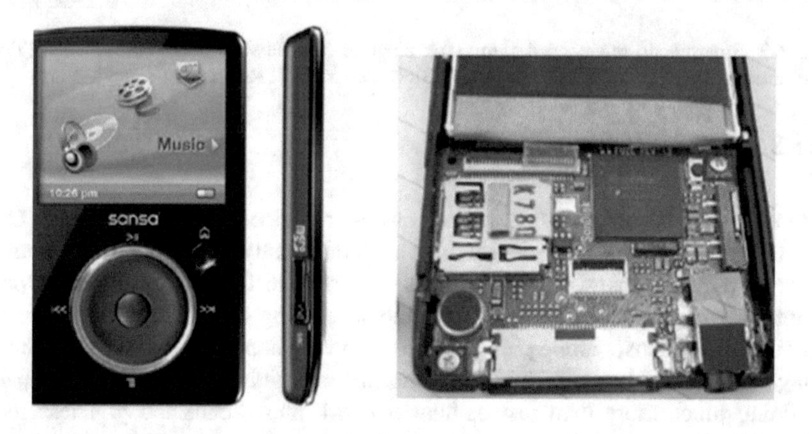

Fig. 6.6 SanDisk Sansa Fuze teardown showing flash memory storage (Courtesy of iFixit)

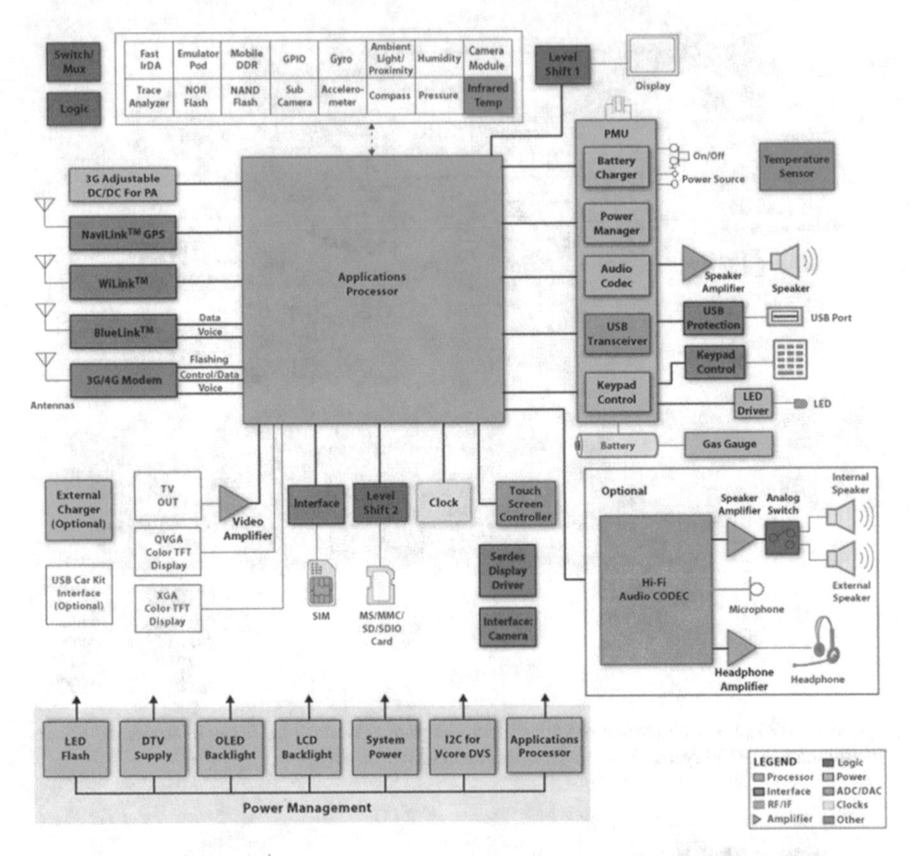

Fig. 6.7 Smartphone reference design block diagram (Courtesy of Texas Instruments) (Texas Instruments, Smart Phone Reference Design Block Diagram, TI Reference Design Library)

6.3.2 *Smartphones*

There is no more ubiquitous portable consumer device than the cell phone. Over two billion of these versatile devices are shipped annually. With such a huge market, there are naturally many market segments with many different capabilities and price points. For many people, particularly in the developing world, their first phone is a mobile phone. Thus, mobile phones represent a very important development in creating personal and business networks. The vast majority of cell phones are capable of doing much more than just cellular network phone calls today. These more advanced phones are called smartphones. In this section, we will focus on smartphones since these require more great storage capacity.

Figure 6.7 shows a smartphone reference design block diagram. According to some reports, in 2016 about 33% of all smartphones were closely based upon reference designs from chip manufacturers.

A modern smartphone provides two-way communications using one of several cellular standards (GSM, CDMA, TDMA, AMPs, etc.). It often integrates one of more of the following functions:

- Vibrating ringer
- Polyphonic ringer
- Touch screen
- Still and/or video camera
- Speakers and headphone jack
- Broadcast radio receiver (FM, AM)
- MP3 player
- PDA functions
- Mobile TV capability
- Various wireless radios
- Location-based services
- Digital storage and storage expansion
- PC connection

Stand-alone music players have been popular devices, although for many people this is a function included on their smartphone, instead of on a separate device. Stand-alone video playout devices have also been popular as separate devices, but today this capability is available in most smartphones and (with their often-larger screens) electronic tablet devices. Smartphones and tablets are available with several operating systems with the most popular being Android, IOS (Apple), and Windows devices.

Smartphones, as well as electronic tablets, often store large amounts of photos, music, or video. As a consequence, they have significant storage capacity needs. In the past smartphones used very small form factor hard disk drives as well as flash memory. Today the mass storage in all smartphones and electronic tablets is flash memory.

Figure 6.8 shows a teardown of an Apple iPhone 7 smartphone

6.3.3 Electronic Tablets

Electronic tablets (like today's smartphones) are often app-based devices with downloadable software applications that can be easily installed and removed. There are also tablet computers that generally have higher-powered processors, as well as more memory and storage capacity, and that have computer operating systems with more permanent software installed. Tablet devices now come in many sizes and capabilities, and it is likely that, for most users, thinner tablet devices will replace traditional laptop computers.

A tablet electronic device is generally capable of playing audio and video content, running games, allowing the creation of text and other types of documents, providing social media tools, and reading and writing email, among other functions. These capabilities are available through downloadable applications, with the application downloaded from a data center (the cloud). To enable these capabilities, the device must have network capability usually available through Wi-Fi and often from cellular networks as well (generally as a paid service). Tablets, along with smartphones,

Fig. 6.8 Teardown images of an iPhone 7 smartphone (Courtesy of iFixit) (iFixit Teardown, iPhone 7, 2016, ifixit.com).

also contain sophisticated environmental sensors, such as accelerometers and GPS radios, that can provide new ways to interact with the machine and location-based services (such as navigation systems).

With increasing bandwidth available through Wi-Fi and cellular networks (4G and soon 5G), the size of content and interactivity available in these devices will increase considerably. This will be important with higher-resolution video, 4 K now and 8 K in the future, and with 360 degree video (which will be particularly important for smartphones and stand-alone VR headsets). In addition, with the improvements in voice recognition, enabled by artificial intelligence capabilities, these devices, along with smartphones, are becoming important devices for automotive entertainment as well as home automation control.

A tablet is thus a sophisticated networked, high-resolution display device that can serve a number of app-based functions. We can see the electronics behind these features in the following figures.

Figure 6.9 shows a block diagram of an electronic tablet reference design. Here are the specifications for this reference design:

- Processor—TI's AM3354 ARM Cortex-A8 microprocessor (up to 1 GHz)
- Operating system—Android ICS 4.0
- System Memory—512 MB DDR3 RAM
- Storage—512 MB NAND Flash and 4GB micro SD card
- Display—7″ WVGA TFT LCD with 5-Touch Capacitive Touch panel
- Audio—Stereo headphone, internal MIC, and internal mono speaker
- Connectivity:
- Wi-Fi 802.11 b/g/n + BT 2.1 module

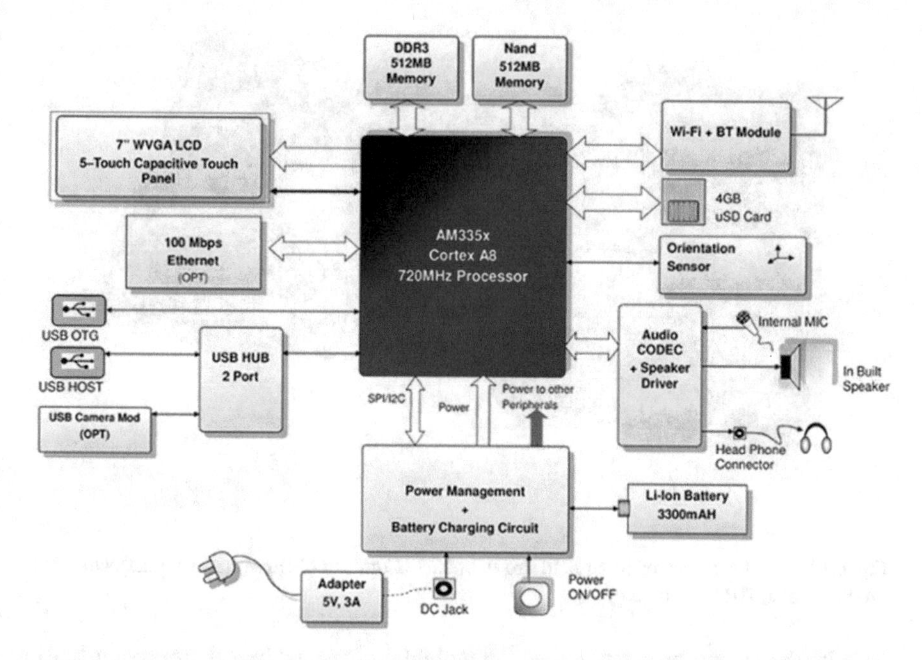

Fig. 6.9 Tablet reference design block diagram (Courtesy of Texas Instruments) (Texas Instruments, E-Tablet Reference Design Block Diagram, TI Reference Design Library)

- Bluetooth 2.1 + EDR
- Ethernet 10/100Mbps (optional)
- Camera—internal USB camera module (optional)
- USB Interfaces—USB high-speed host port and USB OTG (ADB and debug purpose)
- Key/Button—single button for back and home screen, reset switch and power on/off button
- Battery—Li-Ion 3.7 V Nom, 3300mAH with protection circuit
- Power—5 V, 2.5A DC adapter
- Mechanical Casing—black color production quality casing

Figure 6.10 shows images of a teardown of an electronic tablet.

6.3.4 *The Vision of Convergence Devices*

Cell phones differentiate themselves by offering various features in addition to the basic digital cell phone services. Cell phones can include still and video cameras, MP3 players, mobile TV viewers, and PDA capability, including office productivity programs and files. The proliferation of smartphone operating systems that support downloaded applications have turned modern phones and electronic tablets into multipurpose tools for work and play.

Fig. 6.10 Teardown images of an iPad Pro 9.7 tablet (Courtesy of iFixit) (iFixit Teardown, iPad Pro 9.7 Tablet, 2016, ifixit.com)

With the greater processing speed in mobile devices and higher storage capacity (combined with noise cancellation technologies), we have the critical requirements for implanting usable voice recognition into mobile consumer devices. Algorithms based on modern embedded speech recognition have become available on many cell phones. They allow phone dialing by name or by number using your voice and allow voice activation of other cell phone functions. The application understands how names, numbers, and other words sound in a particular language and can match your utterance to a name, number, or command in the phone. Users have found this new functionality straightforward to use and easy to remember.

Voice activation is an important feature for multipurpose devices like cell phones and home control systems. Improvements in artificial intelligence, especially machine learning have made accurate voice recognition and control a reality. In addition to smartphones and tablets, home control devices include voice recognition such as Google Home, Amazon Alexi, and Apple Siri.

You can call a voice-activated assistant on your phone for banking, travel, ordering, and many other activities. The speech recognition algorithms used in these applications are quite large, consuming thousands of MIPs and hundreds of Mbytes and even several Gbytes of memory.

The use of displays that have contrast characteristics more like paper in ambient light would help in viewing content. In 2006, Motorola introduced the 9 mm thick Motofone, which targets emerging markets like India and Brazil. This product uses an electrophoretic display—or EPD—an ultrathin, low-power display often referred to as electronic paper. While the low power of these technologies is appealing, the lack of color and slow page transition prevent users from effective use of popular applications like watching videos. As a consequence, these electronic paper displays have found more use in labeling and active mobile device displays.

Adding additional functions to phones makes these devices more popular and allows offering subscription services that can help retain customers from leaving to competitors and competing technologies. This development combined with the astounding growth of mobile phones of various sorts throughout the world will continue the development and experimentation on using phones as an element in personal networking and device convergence.

Most cell phones today work on cellular networks for their communication and downloading functions and also use Wi-Fi, Bluetooth, and other wireless networking technologies as well. This is perhaps the best example of a mobile convergence device, since the phone can act as a local media and data center for the user and perhaps for other devices that the user has around him. Incorporating IP local wireless networks also opens up the possibility of creating personal area networks (PANs) that allow communication between various devices that a person carries on him/her. Such PANs also open up the possibility of moving some digital storage and other functions off of a single device while still allowing for coordination and synchronization of content when needed. The use of smart watches for fitness monitoring is one example of such distributed functions.

In the network storage chapter, we will explore how a local storage network could be used to provide high-capacity, low-cost storage to such mobile devices without drawing much on their power budgets. A viable storage and communication network and improving local power management and power availability will lead to amazing advances in mobile consumer products such as mobile phones in years to come.

6.3.5 Smart Watches, Jewelry, and Clothing

The technologies that have made devices like the Apple Watch and earlier Android watches possible are processing power, memory/storage, and network connectivity. But today's smart watches are descended from earlier wearable technology, just as devices in the late twentieth century inspired the smartphones and tablets that provide today's ubiquitous portable computing power.

Let's look at a picture of the evolution of electronic watches and see how these devices mutated into today's smart watches Fig. 6.11. In 1972, Hamilton introduced their Pulsar 1 Limited Edition Watch. This electronic watch used off-the-shelf digital logic ICs for timing and displayed time using LED technology. In 1980, Casio introduced their C-80 Calculator Watch which incorporated a very small calculator on the watch using System-on-Chip (SoC) technology—best to use a pen or pencil to operate the keys.

In 1982, Seiko introduced the TV Watch for watching broadcast TV. In 1994, Timex introduced the Timex Data Link 150 Watch that is called the Personal Information Manager (PIM) Watch that could transfer data between a PC and the watch just by holding the watch in front of a computer screen with data on it.

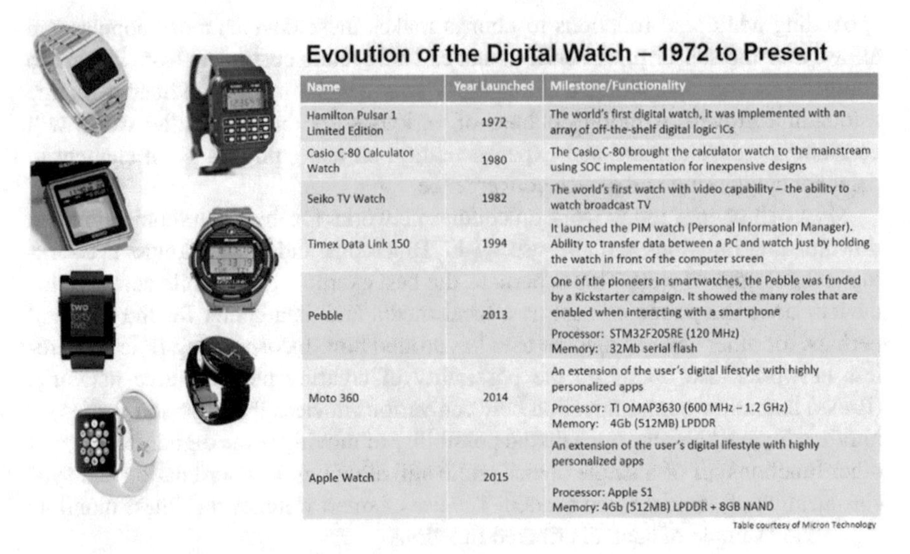

Evolution of the Digital Watch – 1972 to Present

Name	Year Launched	Milestone/Functionality
Hamilton Pulsar 1 Limited Edition	1972	The world's first digital watch, it was implemented with an array of off-the-shelf digital logic ICs
Casio C-80 Calculator Watch	1980	The Casio C-80 brought the calculator watch to the mainstream using SOC implementation for inexpensive designs
Seiko TV Watch	1982	The world's first watch with video capability – the ability to watch broadcast TV
Timex Data Link 150	1994	It launched the PIM watch (Personal Information Manager). Ability to transfer data between a PC and watch just by holding the watch in front of the computer screen
Pebble	2013	One of the pioneers in smartwatches, the Pebble was funded by a Kickstarter campaign. It showed the many roles that are enabled when interacting with a smartphone Processor: STM32F205RE (120 MHz) Memory: 32Mb serial flash
Moto 360	2014	An extension of the user's digital lifestyle with highly personalized apps Processor: TI OMAP3630 (600 MHz – 1.2 GHz) Memory: 4Gb (512MB) LPDDR
Apple Watch	2015	An extension of the user's digital lifestyle with highly personalized apps Processor: Apple S1 Memory: 4Gb (512MB) LPDDR + 8GB NAND

Table courtesy of Micron Technology

Fig. 6.11 The Evolution of the smart watch (The Memory of Watches, Tom Coughlin, Forbes.com blog, April 24, 2015, image courtesy of Micron)

There was some further progress in electronic watches in the 1990s and early 2000s, but the move to today's version of smart wearable devices probably started with the introduction of the Pebble Watch in 2013. Pebble used a 120 MHz processor with 32 Mb of serial flash memory to provide a number of useful functions, including health monitoring. Pebble's initial production was financed by a Kickstarter campaign that allowed early purchase of these useful devices at a significant discount.

Other smart watches and wearable devices followed in quick succession, and by 2014 there were many Pebble-like smart wearable devices in the market. Many of these early devices used a version of the Android operating system. One example is the Moto 360, produced by Motorola in 2014. The 360 and other Android-based watches allowed downloading applications to personalize the wearable experience and provide useful functions. The Moto 360 had a 1.2 GHz processor and 4 GB (512 MB) of LPDDR memory.

In 2015, Apple introduced its Apple Watch, which operates on the IOS operating system and runs a new group of Apple Watch applications. The processor for this device was the Apple S1, and it contains 512 MB of LPDDR3 memory as well as 8 GB of NAND. The Apple Watch set a new standard for wearable technology with its tie into the overall Apple device infrastructure.

Figure 6.12 shows a smart watch reference design block diagram, and Fig. 6.13 shows a teardown of an Apple Watch.

Starting in 2013–2014, some even smaller personal electronic devices started to appear at the CES show. These were smart rings and other forms of smart jewelry. These devices are too small to have a large display or a significant battery, but they were initially used to interface with smartphones. Smart rings are now available that act as fitness trackers, like many smart watches, link people, and control things.

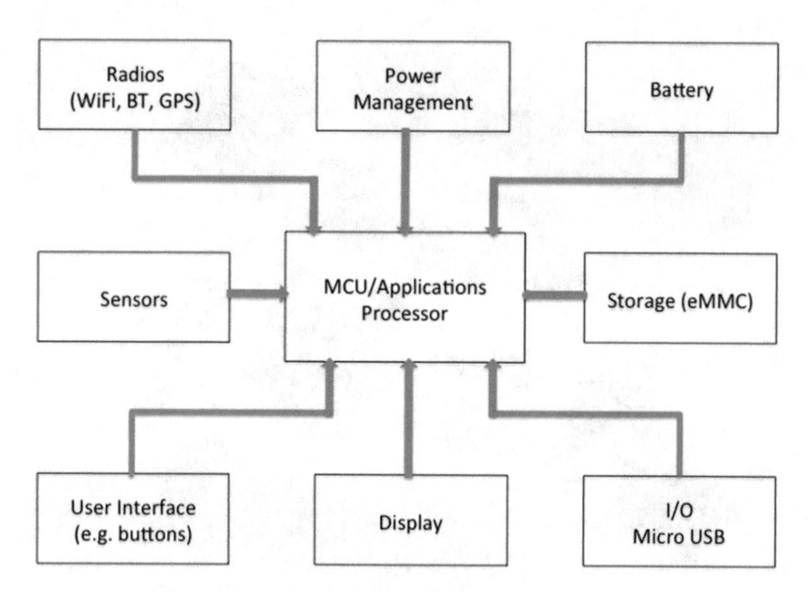

Fig. 6.12 Smart watch reference design block diagram

It is probably a few years before there are high enough energy batteries to make smart rings capable of many more functions, but if voice recognition technology were added, a smart ring could be a very lightweight way to interact with the world. Fig. 6.14 is an example of a smart ring that looks like conventional jewelry.

As batteries get more powerful and as the size of electronics decreases, while the energy efficiency increases, we should see electronics embedded in even more things. This is part of the growth of what is often called the Internet of Things (IoT), smart connected things. There are already early examples of smart clothing with displays and sensors built in. With the design of flexible circuits, smart clothing could become even more complex.

Someday our clothing, our glasses, jewelry, and other common things could replace smartphones, TVs, and many other common entertainment and information devices as these functions are built into things that we wear. Clothing could contain cameras and other sensors to help us interface with and remember the world around us, especially combined with artificial intelligence and local or cloud-based digital storage. Clothing might also become an electronic display device, allowing static or dynamic image displays that could be tuned to how we want to appear to the world or to our innermost feelings. There are efforts to implant devices in humans for healthcare and other purposes, and these devices would naturally interface with a PAN that included our smart clothing.

When everything can be smart, what will enable us to do? How will we entertain ourselves, educate ourselves, and interact with other electronic-enhanced people? How will we retain our privacy? It will be an exciting new world.

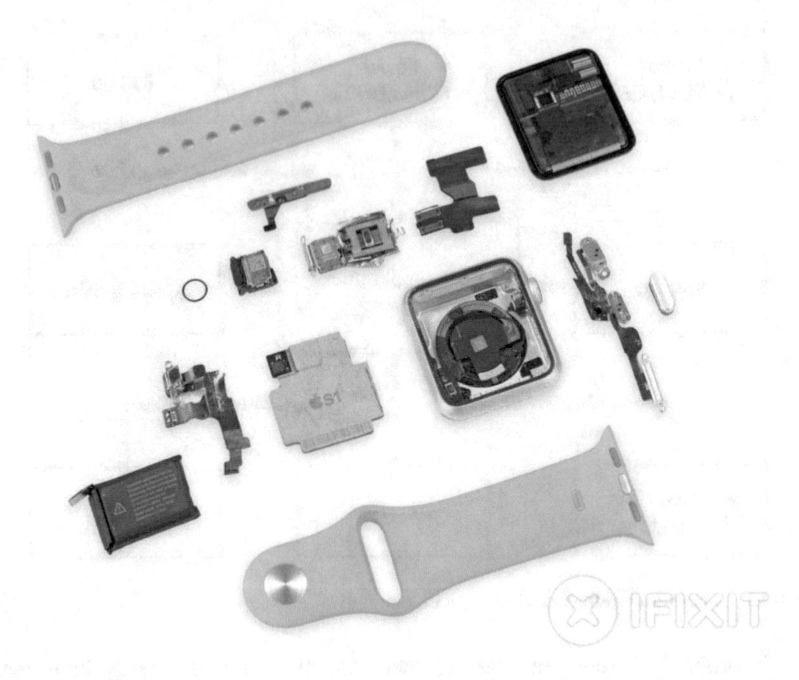

Fig. 6.13 Teardown images of an Apple Watch (Courtesy of iFixit) (iFixit Teardown, Apple Watch, 2015, ifixit.com)

Fig. 6.14 Smart ring (Ringly Smart Ring, https://www.wareable.com/meet-the-boss/ringly-ceo-christina-mercando-smart-jewellery-2016)

6.4 Cameras and Camcorders

Stand-alone digital still cameras used to be the only way to take a really good picture. But with the rise of high-resolution cameras in smartphones, many people don't even use stand-alone still or video cameras. However, for the best images, high-end digital still and video cameras offer larger lenses, filters, and other features seldom found on smartphones.

Stand-alone digital still cameras mostly use flash cards of various formats to store images. Very high pixel count cameras used by professional photographers used to use removable 1 inch HDDs in a compact flash form factor, but with the rise of high-capacity flash cards, this is no longer the case.

The stand-alone tape camcorder market is not very large and has been declining since 2010. This small market penetration is probably due to the $300–2000 price tag that many of these devices command. This is too high for most consumers. These high prices are driven in part by the recorder's very intricate recording mechanism.

For a while, HDD-based camcorders had the highest storage capacity, but with higher capacity flash memory at affordable prices, camcorders have moved to flash memory. Likewise, optical disk-based camcorders are a vanished species.

Flash card-based camcorders are available for less than $100 from some Chinese manufacturers. Better known brands with better optics and optical sensors can cost $200 or more, sometimes much more. Like with stand-alone still image cameras, those seeking higher quality than they can get on their smartphone will generally pay for the more expensive features.

6.4.1 Layout of a Digital Still Camera

Figure 6.15 shows a block diagram for a digital still camera. These cameras, although often called still cameras, can usually take videos as well as still images. In fact, there are high-end DSLR cameras that are routinely used for professional video work. With the improved optical quality of images shot from smartphones, stand-alone digital cameras are generally used to take higher-quality images with better lenses and optics than those available on phones. These features add to the cost of the camera.

As a consequence, the low-end still camera market is pretty much dead.

Figure 6.16 shows a teardown of a digital still camera showing various labeled components.

The digital camera uses an image sensor (CCD or CMOS) to convert light directly into a series of pixel values that represent the image. The major components of a digital still camera are:

- The CCD or CMOS image sensor converts light photons into electrons at the photosensitive sites in the CCD or CMOS image sensors.

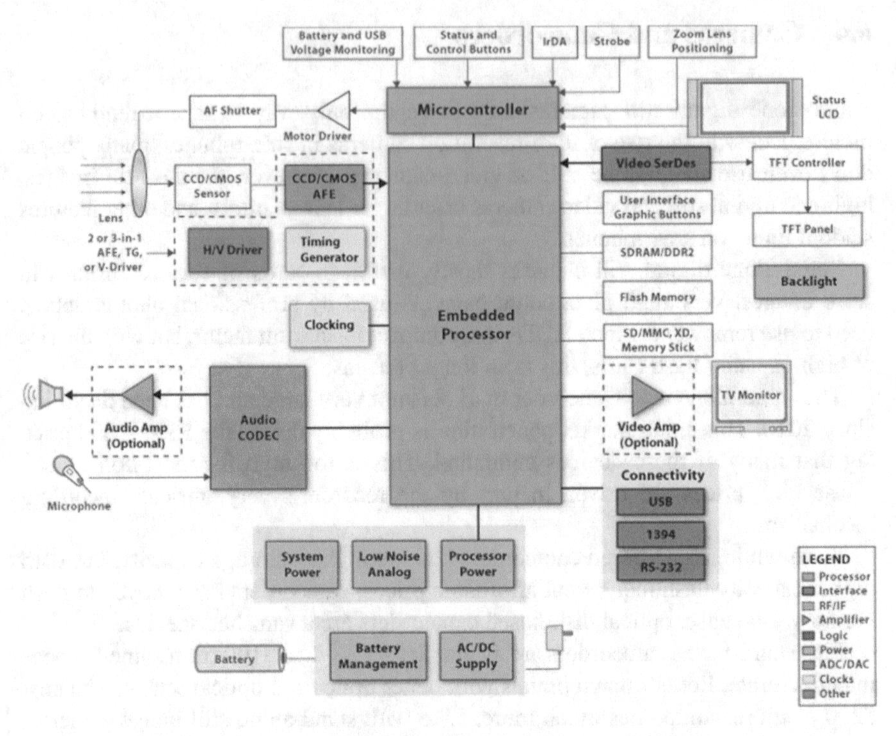

Fig. 6.15 Digital still camera block diagram (Courtesy of Texas Instruments) (Texas Instruments, Digital Still Camera Reference Design Block Diagram, TI Reference Design Library)

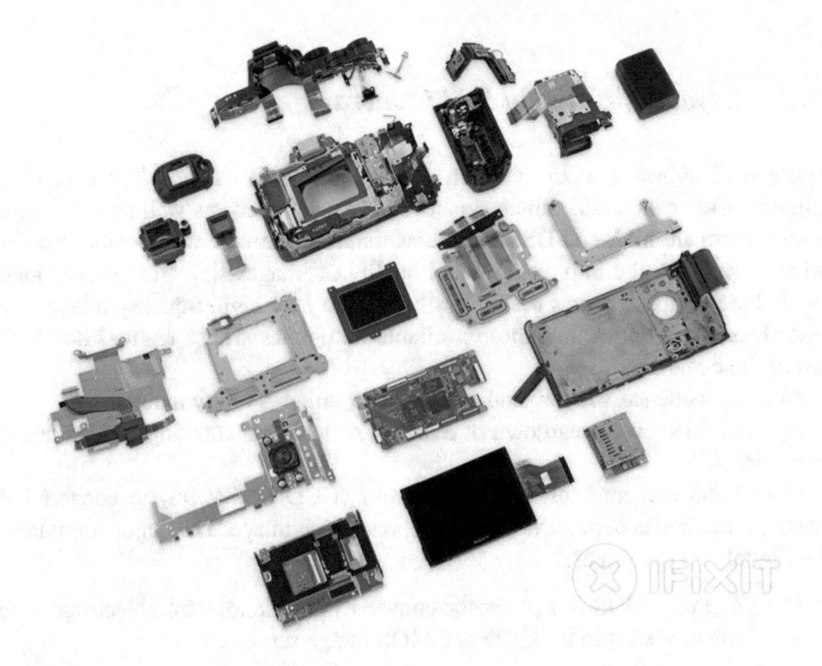

Fig. 6.16 Teardown image of digital still camera (Courtesy of iFixit) (iFixit Teardown, Sony A7R II, 2015, ifixit.com)

- The front-end processor filters, amplifies, and digitizes the analog signal from the image sensor using high-speed analog to digital conversion.
- The digital image processor handles industry-standard computationally intensive imaging, audio, and video algorithms. It also controls the timing relationship of the vertical/horizontal reference signals and the pixel clock.
- The various memory stores executing code and image data files.
- The audio codec performs digital audio recording/playback under the control of the DSP.
- The LCD controller receives digital images from the camera and images them on the LCD.
- The power conversion electronics converts input power from AC or USB to charge the camera battery and control the power management functional blocks.

6.4.2 Layout of a Digital Video Camera

Digital video cameras are often referred to as camcorders. These devices have used small form factor digital videotape, optical disks, and hard disk drives for capturing the video images for many years. Today the most common camcorders use flash memory, often in removable card formats.

Figure 6.17 is a block diagram of a digital camcorder showing the basic design elements in a digital camcorder. Note that except for the addition of video content and the larger storage capacity required for video content, there are many similarities to still cameras.

Figure 6.18 shows a teardown of a digital still camera showing various components.

6.4.3 Storage Requirements for Digital Camcorders

Many modern digital video recorders allow capture of digital still images as well as video images, and the opposite is true of today's still image cameras. In many regards, a still picture is just a single frame of motion for today's equipment.

The resolution of video as well as optical content in camcorders will continue to grow, driven by HDR images and especially 360 stereoscopic videos. Digital still image megapixel resolution has been doubling about every 4 years. With the growth of high-definition video in the home, we can expect that for the next several years, video resolution in camcorders/cameras will on average double about every 4 years as well.

Smartphones and perhaps smart clothing may be the way we capture most images in the future. However, large lenses with significant light-gathering capability and high-quality lenses provide features in most expensive stand-alone cameras that can't be matched with smartphones, yet.

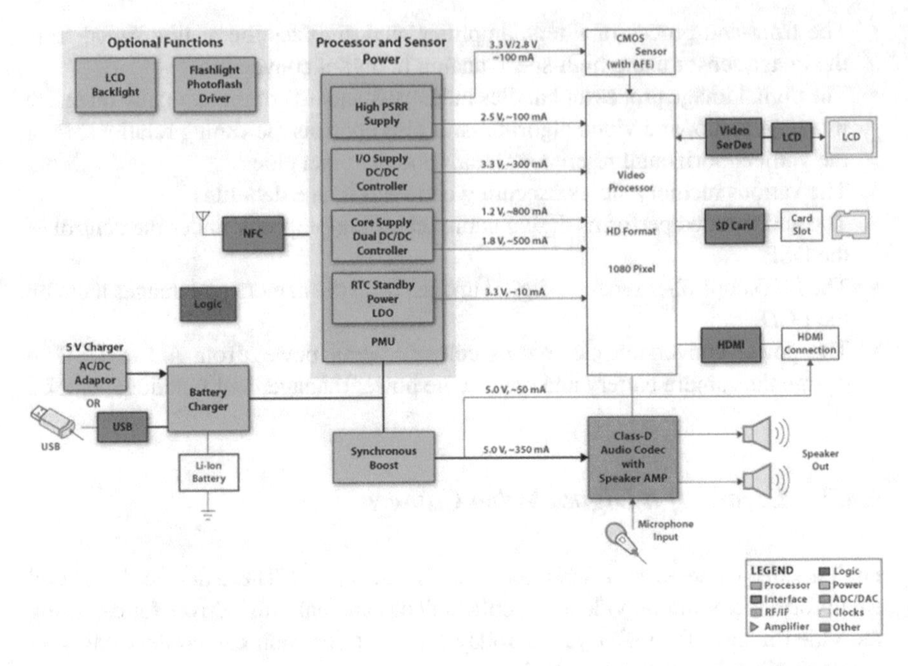

Fig. 6.17 Block diagram of a digital camcorder (Courtesy of Texas Instruments) (Texas Instruments, Digital Video Camera Reference Design Block Diagram, TI Reference Design Library)

Fig. 6.18 Teardown image of GoPro Hero 4 video camera (Courtesy of iFixit) (iFixit Teardown, GoPro Hero4 Camera, 2015, ifixit.com)

6.4.4 Road Maps for Camcorder Digital Storage

As the resolution of integrated (such as smartphones) or stand-alone cameras increases, the price drops and more people start to share their content, greater digital storage capacity will be required, and greater network bandwidth will be required to share this content. More cameras are equipped with wireless capability, making it easy to transfer content to a nearby computer on a local network, to cloud storage facilities, and possibly even to backup images and video to a wireless mobile storage device that you carry with you on your *personal area network* (PAN).

Developments in camera optics are also in for major changes once stodgy optical technology is transformed. Researchers are developing new devices using negative index meta-materials to create more complex and compact optical paths allowing even more powerful and compact imaging systems.

Multiple wavelength recording (not just visible optical wavelengths) is possible in a device that can be easily incorporated into other devices or turned into a lightweight and compact stand-alone imaging device. This multiwavelength imager would allow the capture of much more than just surface visible light images around us (imaging combining visible light, infrared light, and even higher or lower frequencies).

Very small laser projection technologies could help turn digital cameras into their own projectors for playing back their content, or viewers could see content projected directly into their eye using augmented reality. All of these capabilities will encourage people to take more images and store more of those images in more places.

The increasing ubiquity of still and video cameras, GPS location devices, and other technologies, combined with artificial intelligence to connect and make sense of these images, could lead to a new type of device discussed latter in this section that allows the recording of events as they happen in a person's life and storing and even sharing parts of the content with others.

The visual image is one of the most powerful ways for humans to communicate with each other and to learn about the world around us. It is only natural that we should quickly incorporate these new advances in image capture, processing, and presentation to enhance the human experience and capture it for future generations.

6.5 Other Consumer Devices

There are several other types of consumer devices that utilize digital storage to enable their operation. We shall briefly discuss some of these as well as a short section speculating about future products enabled by very large mobile digital storage.

6.5.1 Mobile Game Systems

Mobile games tend to use semiconductor memory for game storage. The graphical requirements for the smaller screens do not require as large storage files as do home-based systems, and so lower-capacity flash memory or other nonvolatile solid-state memory can be used. The use of solid-state memory in this application is wise since these mobile game systems often get rather harsh treatment by children playing with them.

6.5.2 Handheld Navigation Devices

Handheld navigation devices run a gamut from inexpensive systems selling for less than $50 to more elaborate systems that can cost several hundred dollars. GPS-based navigation systems are used for automobile and marine navigation as well as for handheld use while hiking and doing other outdoor activities. In addition to stand-alone devices, GPS location functionality is often incorporated with other consumer products, such as smartphones.

A handheld GPS navigation system reference design schematic is shown in Fig. 6.19.[3] Except for the battery management and incorporation of display and other functions in the infotainment system, a GPS navigation system in an automobile will be similar.

The Global Positioning System (GPS) works on the principle that if you know your distance from several known locations, then you can calculate your location. The known locations are the 24 satellites located in 6 orbital planes at an altitude of 20,200 Km. These satellites circle the Earth every 12 hours and broadcast a data stream at the primary frequency L1 of 1.575GHz that carries the coarse-acquisition (C/A) encoded signal to the ground. The GPS receiver measures the time of arrival of the C/A code to a fraction of a millisecond and thus determines the distance to the satellite.

The major components of a GPS receiver are:

- Front end—the GPS L1 signals (maximum = 24 signals) at 1.575GHz are received at the antenna and amplified by the low-noise amplifier (LNA). The RF front-end filters, mixes, and amplifies using analog gain control (AGC) and shifts the signal down to the IF frequency where it is digitally sampled by an analog to digital converter (ADC).
- Baseband processor/CPU—the ADC samples of GPS C/A code signals are correlated by the digital signal processor (DSP) and then formulated to make range measurements to the GPS satellites. The DSP is interfaced with a general-purpose CPU, which handles tracking channels and controls user interfaces. TI OMAP integrates both DSP and ARM processor on the same chip.

[3] Texas Instruments, GPS: Personal Navigation Device Block Diagram, TI Reference Design Library.

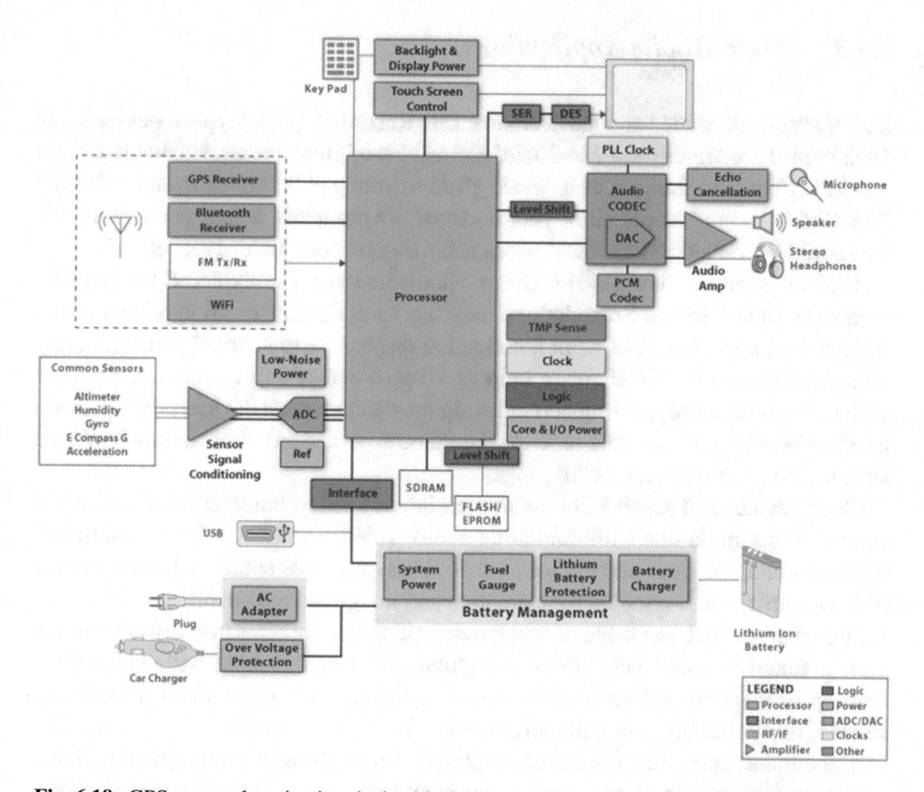

Fig. 6.19 GPS personal navigation device block diagram (Courtesy of Texas Instruments)

- Memory—the processor runs applications stored in memory. The OS is stored in nonvolatile memory such as EE/FLASH/ROM. Applications may be loaded in NAND FLASH or DRAM.
- User interface—allows user to input/output data from the receiver using input commands via microphone, touch screen, and output MP3 to the earplug.
- Connectivity—allows the receiver to connect to the USB port.
- Power conversion—converts input power (battery or wall plug) to run various functional blocks.

As GPS receivers have declined in size and decreased in cost, they have become a standard function built into many consumer devices. Today location-based services using GPS can be found in automobiles, cameras, phones, smart watches, and tablet electronic devices and will be in many more products over time.

Regarding digital storage requirements for such devices, the storage needs are similar to those discussed in the section of this chapter on automobile navigation and entertainment systems. As the map details become more precise, the digital storage requirements increase. Also, if there is a need for real-time data updates to maps, terrain, or road conditions, a rewritable device such as flash memory may be desirable.

6.5.3 Other Mobile Applications

Don Norman speculated about a "Personal Life Recorder" (PLR) type of device in his 1992 book *Turn Signals Are The Facial Expression of Automobiles*. He theorized that these PLR's would start out as a device given to young children, called the "Teddy." The "Teddy" would record all of your personal life moments, and as you mature, the data could be transferred to new devices that matched your maturity level.

Projects such as "MyLifeBits" from Microsoft have also explored the requirements for such devices. Someday we may use such devices or an application that collects and analyzes input from a collection of sources to share clips of our lives with our families or friends or to help recall past events and contacts. Combined with a capability of organizing and indexing the content, such devices could provide a very powerful personal data base that could be accessed any time. In this book, we shall refer to such devices as "life logs."

Such devices and capabilities are in their infancy today, but in practice one could capture and sample one's life including audio, video, and GPS information linked together collected from one device or more likely multiple sources. The pieces are now available, and it remains for a creative and imaginative company to find a market for such product. In an age of social networking and cameras in every cell phone, such a function could become very popular. With volume and with technology development incorporating mobile recording devices and other input into such a service, this could become quite affordable.

Depending upon the resolution captured, these devices would require huge amounts of information, and the support system to store and organize the total aggregate content in the home or in the cloud could be even greater; a terabyte in the pocket and a petabyte in the home would not be out of the question. The proliferation of such services could lead to the single biggest use of digital storage in the world, since there are a lot of people out there and thus a lot of people that may want to record their lives.

6.6 Chapter Summary

- Automobiles will use digital storage for entertainment and navigation purposes. These devices must be able to withstand very harsh environments and yet continue to function. As the complexity of automobile navigation aids and automobile entertainment resolution increases, the storage capacity requirements will increase. Automobiles will also be part of a home network, allowing them to obtain and share content with that network. This could significantly automate the acquisition of content and navigation information.
- Music and video player technology continues to develop. For highly compressed content (such as MP3 music), flash memory will probably provide all the digital storage needed in the future. However, for less compressed (and higher-quality) audio as well as video, there may be a need for higher-capacity, lower-cost storage, such as hard disk drives.

- The use of a larger screen on video players and the increased resolution of the content favor the use of HDDs for low-cost storage for applications such as automobiles, at least until the cost of enough flash drops low enough.
- New power source and wireless power technologies could provide extended usage life for mobile devices and enable the use of technologies that require more power such as small projector systems for projecting images and video or radios to stream content to augmented reality or virtual reality systems.
- Digital still and video cameras are being incorporated into many other devices as well as continuing to exist as a stand-alone device category. Stand-alone devices favor higher-resolution, better image sensors, and better optics and serve a higher end of the market.
- For many folks, their smartphone cameras are sufficient for many of their needs. These products will require larger storage within as content resolution increases and also require larger storage capacities for off-loaded content—such as into home storage, direct or network attached storage devices, and the cloud.
- Ultimately the growth in available capacity of mobile digital storage components combined with cloud-based services to integrate and analyze information from multiple sources will enable the creation of devices that can record a person's life experiences as they happen. Such a "life log" will require significant digital storage capacity in the device as well as in the home network.

Chapter 7
Developments in Mobile Consumer Electronic Enabling Technologies

7.1 Objectives in this Chapter

- Determine what sort of display technology is useful in a consumer device.
- Explore the trade-offs in performance and power for mobile music and video players and how new energy sources could lead to richer content available for viewing using these devices in the future as well as longer time between charges.
- Lay out a discussion for consumer metadata standards and how these could be used by cloud and on-premise artificial intelligence.
- Look at how voice and image recognition will change our interaction with our devices and create useful metadata.

7.2 Display Technologies in Mobile Devices

Creating inexpensive and easy to use touch-based high-resolution display technologies has been a key development in mobile electronic devices. We expect that voice-based interfaces will create new ways to interface with our consumer devices. Let us look at the future developments in video display and voice recognition technology in consumer devices.

7.2.1 Mobile Device Displays

Today people watch content on many different size displays, from large theater screens to home displays to tablets and smart phones. All these different types of displays require content that is tailored to them. Real-time transcoding of content to

T.M. Coughlin, *Digital Storage in Consumer Electronics*,
https://doi.org/10.1007/978-3-319-69907-3_7

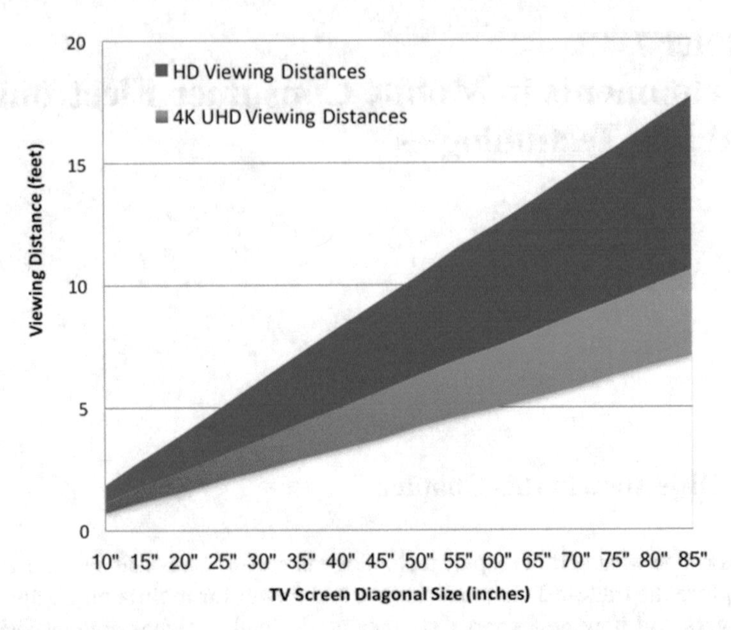

Fig. 7.1 Comparison of optimal viewing distance for HD and 4 K UDH versus display diagonal size

match the needs of a particular communication channel and different screen sizes provides important efficiencies, as these different content formats can be created more or less as needed (and thus not stored before use).

Higher definition is generally most noticeable for very large screens or where the viewer is liable to be very close to a screen. Because the eyes of computer and tablet users are generally close to their screens, some computers and tablets have 4 K resolution screens, and users can see the difference.

For home viewing of a TV with comfortable viewing distances (generally more than 5 feet from the screen), the screen size needs to be more than 60 inches diagonal for the benefits of 4 K resolution to be apparent. It seems that higher resolution has the greatest value for smaller screens where the viewer is close, as well as very large screens where the viewer is some distance away. Figure 7.1 shows a plot of the optimal viewing distances for HD as well as 4 K UDH content. For 360° video, even 8 K resolution per eye can create a greater sense of immersion, meaning that the complete video would be 16 K resolution. Higher resolution means higher bandwidth for moving the content and higher storage capacities to store it.

While resolution perception is a function of the size of the screen and the distance of the viewer, perceptions of color and high dynamic range seem to make a difference to viewers for all size screens and reasonable distances from a screen. Let's look at the way modern display devices provide colors and dynamic range.

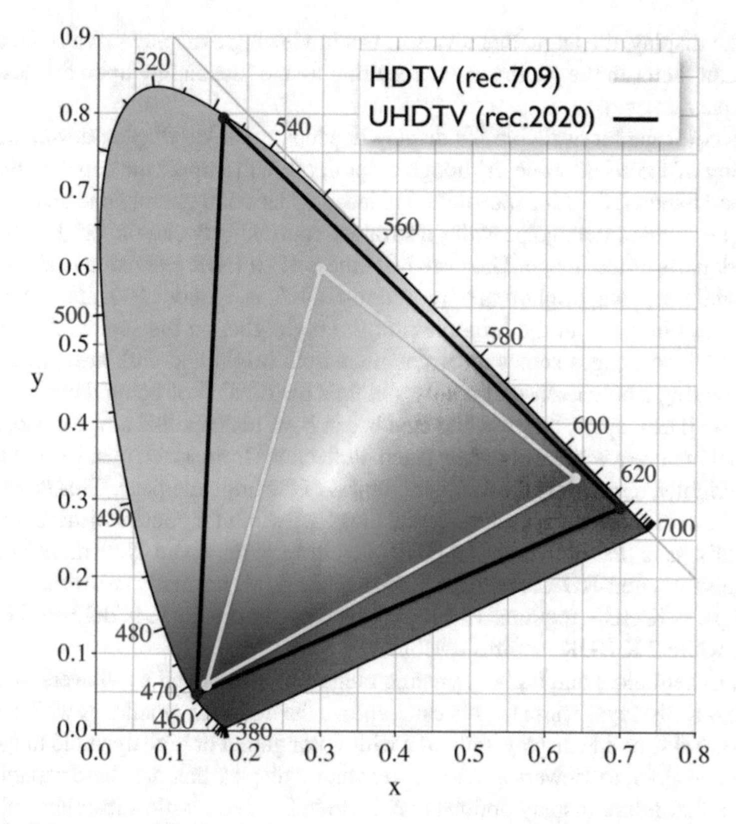

Fig. 7.2 Color spaces of display devices overlaid on the CIE 1931 color space

7.2.2 Color

In 1931 the International Commission on Illumination (CIE) defined a quantitative relationship between physical colors (light wavelengths) and human color vision. This resulted in a chromaticity diagram that can be used to specify how the human eye will experience light with a given wavelength as a particular color. This relationship defines the resulting color space.

Figure 7.2 shows the CIE 1931 color space as well as the UHDTV (Rec 2020) color space in the outer triangle as well as the HDTV (Rec 709) color space in the inner triangle. The x- and y-axis are derived from the brightness and chromaticity of each color. The illuminant D65 point is defined as white in both of these display color spaces. The numbers on the outside of the locus of the CIE 1931 color space are the wavelength of the light corresponding to that color.

Rec 2020 has a wider color gamut than Rec 709 and thus would show more saturated and richer colors. Some types of displays can show a greater gamut of colors within the defined color space. OLED displays as well as quantum dot displays provide richer colors than most LED displays. Note that, in addition to distance from the display, the angle that a viewer has in viewing a display can also be a very important factor in the quality of the resulting image (depending upon the design of the display device).

The color gamut available for display depends upon the display device and the encoding of the color space. Although color depth will impact the capacity requirement and bandwidth of the transmitted video, a wider color gamut generally will not.

Higher dynamic range provides a stronger contrast between the bright parts and the dark parts of the image. The peak brightness of an HDR television is about 1000 nits,[1] while the peak brightness of most non-HDR TVs is about 100 nits. Thus, there is a tenfold increase in the brightness of the highlights on the screen. As a result, HDR in video images shows more details in both bright and dark areas of the display, creating a better sense of reality and thus the illusion of being there.

In an HDR display, bright white clouds can have textures like a real cloud, rather than looking like a washed out white patch. Reflections from water, metal, and glass are much brighter, resembling the actual brightness of the original image. Greater dynamic range requires more storage capacity and thus bandwidth for content distribution.

Netflix says that delivering an HDR picture requires about 20% more bits than the equivalent non-HDR resolution.[2] Thus, while 4 k is normally delivered by about 15 Mbps, 4 K HDR requires 18 Mbps. Likewise, normal 2 K is delivered by 5–6 Mbps, while 2 K HDR requires 8 Mbps.

So, richer colors and higher dynamic range can be enjoyed by viewers in 2 K as well as 4 K displays. These factors can improve the image on smaller as well as larger screens. A 4 K or 8 K display without a wide color gamut or high dynamic range may not look as good to viewers as a lower-resolution display that has these capabilities.

Note that future display options may include flexible plastic substrates and even small projection devices built into consumer devices. Only miniaturization, energy management tools, and longer lasting mobile power sources are needed to enable longer lasting mobile consumer experiences.

7.3 Mobile Power[3]

In mobile devices, power is a serious design parameter. All the components in a mobile device require power. The power requirements for different system components (after evaluating and optimizing performance and usage life trade-offs for the expected product applications) can be used to create a power budget, which then defines the mobile power supply needs of the device.

[1] A nit is a unit of visible-light intensity. One nit is equivalent to one candela per square meter. One candela is approximately the amount of light emitted by a common tallow candle.

[2] Netflix: High dynamic range is “more important” than 4 K, Sophia Curtis, The Telegraph, Jan. 12, 2015.

[3] Much of this section is derived from an IEEE Consumer Electronic Society Future Directions White Paper, *Safe Advanced Mobile Power*, February 2015.

An important way to control device power usage is with buffering and caching and power off modes when a device is inactive. Enormous advances have been made in making the electronics more efficient, even with greater loads, but in the end, better power sources are needed. Let us look at ways that power can be minimized and then at the potential for higher mobile power devices.

The raw data rate of a storage device such as a hard disk drive (or a flash memory device) is much greater than required by an MP3 or MPEG player application. A buffer is used as a host speed matching device. The size of the host buffer will determine the power dissipation; more buffer means less power-hungry spin up/down drive cycles (for a HDD), resulting in lower average power dissipation. Also for online streaming, more buffering allows continuous streaming even when there are temporary network glitches.

Trade-offs in buffer size, power, and cost help host manufacturers select a buffer that meets the power dissipation goals of the application while minimizing cost. Typical music and video player buffers are in the range of 8–16 MB. These buffers can easily match the streaming requirements for video and audio content.

Battery technology is advancing at a gradual rate, estimated as roughly doubling in energy density with every decade in time. Based upon this estimate for battery energy density growth, it will take decades to develop a battery that can supply enough energy to power a smart phone for a week's worth of "normal" use. In order to accelerate this development, we need something like a "Moore's law for mobile energy."

Although popular mobile devices use energy storage devices (batteries) as their power source, battery technology for consumer mobile applications has not changed much since mobile phones were introduced to mainstream consumers in the 1990s. Advances in mobile battery technology are much slower than increases in processing power, network speeds, and digital storage technology.

Advances in novel battery technologies will enhance the IoT by enabling a plethora of new applications. Features that are "always on" are enabled by higher energy mobile power sources. These advances will revolutionize the IoT's ability to interact with its surroundings and provide a virtual extension of human knowledge and influence in ways we do not yet fully appreciate. Likewise, the availability of more mobile device power will enable the operation of multiple simultaneous radios and widespread use of technologies such as microprojectors in common mobile devices.

There are a few ways to estimate the required energy for a week's worth of activity. One approach is to look at the current energy stored in a phone and make an estimate based off that number about how much energy is needed for a week. Another approach is to make estimates of actual energy used for various discrete operations and then calculate how much time and thus energy is spent on these operations in the course of a regular day.[4] Of course, we also need to take into account that the cell phone might have more use during weekdays than weekends. However, all of these estimates depend upon some knowledge of user patterns that may not be valid for all individuals.

[4] P. Perrucci, F.H. P. Fitzek and J. Widmer, Survey on Energy Consumption Entities on the Smartphone Platform, IEEE Vehicular Technology Conference (VTC Spring), 2011

Table 7.1 Rated and actual energy use and battery energy storage for conventional smart phones

Product name	Mfg. standby		Vendor reported run time	Actual run time	Installed battery
1	HTC dream	406 h	5 h 20 min	2 ~ 3.5 h	Li-ion 1500 mAh
2	Google nexus one	290 h	7 h	3.5 ~ 5.5 h	Li-ion 1400 mAh
3	Apple IPhone 5	225 h	8 h	3 ~ 5 h	Li-Po 1440 mAh
4	Samsung galaxy 5	375 h	6.45 h	2.5 ~ 4.5 h	2800 mAh
5	Nokia lumen 1520	32 days	24 h	9 ~ 10.8 h	3400 mAh

Let's look at the energy in the batteries of conventional smart cell phones. Table 7.1 shows examples of mobile smart phone products (from 2015), their current consumption, and their estimated run time on the installed battery.

Item 5 seems to be an outlier on actual run time. Item 5 is a cell phone with a much higher battery capacity, which is a potential solution, although with conventional technology adds cost, size, and weight. All these devices tout being able to "run" for 34 h, but in reality, the battery charge tends to run out in 2–8 h, depending on use. There are many factors for this, including the applications being used as well as the usage environment and features enabled at a given point in time. In addition to the applications explicitly in use, services running in the background may also consume power.

In order to make some calculations on the amount of energy that is required for a week's use, let us look at a model of a casual and power cell phone user. The casual user might get 5 h of use with a 1400 mAh (1.4 Ah) battery. If this was consistent across 7 days, the total energy requirement is 7 X 1.4 Ah or 9.8 Ah. Assuming this is a 3.7 V device, then the actual energy used is 9.8 Ah X 3.7 V = 36.3 Wh or 130,536 Joules.

On the other hand, a power user might want 12 h of intense use in a day with lots of radios running (Bluetooth, Wi-Fi, GPS, and NFC), downloading lots of data and watching movies. This user may consumer 1400 mAh (1.4 Ah) in only 3 h and actually want to consume four times more power in the day or 4 X 1.4 Ah = 5.6 Ah. If this is a 3.7 V device, the energy required for 1 day would be 5.6 Ah X 3.7 V = 20.7 Wh. A 7-day energy requirement for such a user would then be 7 X 20.7 Wh = 145 Wh = 522,144 Joules.

Today's cell phone batteries typically have about 1400 mAh = 1.4 Ah of power or 1.4 Ah X 3.7 V = 5.2 Wh of power. Based upon our calculations, a casual phone user may need about 7 X more energy for a full week, while a power user may need about 28 X more energy that these batteries.

Thus, the required energy for a week's cell phone use could range from about 36.3 Wh (130,536 Joules) to 145 Wh (522,144 Joules). This is well beyond the capability of conventional batteries used in cell phones. In order to create an energy

Table 7.2 Important characteristics for mobile consumer electronic devices

Condition	Requirement	Units	Comment
Product life time	3	Years	Consumer environment
Operating temperature normal use	0–40	°C	Indoors
Short-term operating temp	40–60	°C	Use on beach or in car in hot climate
Min/max operating humidity	15–80	%Rh	
Power consumption			Target user models but also includes standby
Reliability	Not-in use—Class 1.1		Weather protected, partly temperate controlled (represents home/office conditions)
	In-use—Class 7.3E		Partly weather protected and partly use in non-weather protected (indoor/outdoor usage)
Storage temp range	−10 to 45	°C	
Storage humidity range	10–90	%HR	
Direct sunlight exp	600	h	
Packaging			Class 2.2—Shock/bump/vibration
Thermal expansion			Not to produce permanent bow or breakage
Safety and flammability requirements			Meet UL-94 to V0 level

source that can meet these challenging requirements, we will need to go beyond the capabilities of currently used technologies in mobile devices.

A candidate power source must meet various environmental and safety specifications. For instance, nonoperating temperatures specified for mobile phones are in the range of −10° to 45 °C. Operating temperatures are generally in the range of 0–40 °C. Relative humidity should be in the range of 5% to 95% noncondensing. Maximum operating altitude is 3000 m.

A more complete list of characteristics and product requirements for mobile consumer electronics devices is shown in Table 7.2.

Table 7.3 shows a typical Li-ion cell's specifications and the safety standards that it must pass for use in consumer products.

Some of the current base standards that contain relevant safety requirements are cited below. Specific requirements for the cited standards (limits, test criteria, etc.) are cited in detail in each standard. The actual use environment, energy sources (types and levels) present in the product will determine which requirements, and tests are applicable to each product.

Table 7.3 Typical Li-ion battery cell specifications

Item	Specifications
Capacity	1800 mHA
Impedance	<180 mohm
Dimensions	4.8 mm × 50 mm × 67 mm max
Weight	50 g
Maximum charge voltage (V)	4.28 V
Maximum charge current (I)	1CmA
Minimum discharge voltage (V)	3.0 V
Maximum discharge current (I)	1CmA
Charge temperature	0 °C to +45 °C
Discharge temperature	−20 °C to +60 °C
Storage temperature	−5 C to 25 + −5 C
Storage humidity	≤75% Rh
Charge voltage	4.2 V
Nominal voltage	3.7 V
Nominal capacity	1800 mAh 0.2 C discharge
Charge current	Std 0.2 C; rapid charge 1 C
Std charging method	0.2 C contrast charge t0 4.2 V than constant voltage 4.2 V till charge current decline to ≤0.02 C
Charging time	3 h Std; rapid 2 h
Maximum charging current	1 C
Maximum discharge current	1.5 C
Discharge cut-off voltage	Change cut-off voltage 3.0 V
Operating temperature	Charging 0C-45 C; discharging -20 C-60 C
Storage temperature	−5 C – +35 C
Cycle life	≥ 300 cycles to 80% capacity based on charging condition
Self-discharge	Using Std charging conditions, 90% capacity after 28 days when measured using 0.2C discharge
Initial impedance of the cell	≤60 m ohm internal resistance at 1KHz after 50% charge
Temp characteristics	Capacity comparison at various temperatures measured with constant discharge of 0.2C with 3.0 V cut-off when compared to 100% capacity at 25C, −20C at 60%, 100% at 25C, 85% at 60C

(continued)

Table 7.3 (continued)

Item	Specifications
Vibration test	After Std charge, expose to vibration between 1 Hz and 55 Hz with 1 Hz variation/min. The excursion of vibration is 1.6 mm for exposure of 30 min per axis (*XYZ*)
Drop test	Drop test 1 m on concrete floor—Check for leak or fire
Special use instructions	Visual checks for cracks, scratches, and leakage
Guarantee	Supplier dependent
Regulatory tests passed	Must pass all safety tests

7.3.1 Safety

(a) Basic safety—consumer technology, medical, security, and signaling devices—base standards, IEC 62368–1 Ed. 2, IEC 60950–1 Ed 2, A1/A2, IEC60601 Ed. 3 (may include risk management), and others, addressing:

 (i) Energy sources
 (ii) Electrically caused injury
 (iii) Electrically caused fire
 (iv) Hazardous substances
 (v) Mechanical injury
 (vi) Thermal burn
 (vii) Radiation
 (viii) Functional safety

(b) Battery–primary and secondary–IEC60086, IEC61960, IEC62133, UL1642, UL2054
(c) Optical radiation–lasers and LEDs—IEC60825–1, IEC62471, FDA 21 CFR 1010 and 1040
(d) Medical—software—IEC 62304; quality systems, ISO13485
(e) RF exposure—specific absorption rate (SAR)—IEC 62209–1, IEC 62209–2, IEEE 1528

7.3.2 Other Requirements

(f) Electromagnetic compatibility (EMC)—Federal Communication Commission (FCC), Industry Canada (IC) and European Union (EU) EMC Directive, etc.
(g) Wireless regulatory—FCC, IC, EU Radio and Telecommunication Terminal Equipment (R&TTE) Directive, etc.
(h) Wireless interoperability—Bluetooth certification, Wi-Fi Alliance, etc.
(i) Sustainability—impact on environment during product life cycle
(j) Responsible sourcing—deals with human rights, working conditions/hours, social auditing, supply chain risk management, etc.

(k) Chemicals—Restriction of Hazardous Substances (RoHS); Waste Electrical and Electronic Equipment Directive (WEEE); Registration, Evaluation, Authorization and Restriction of Chemicals (REACH); polycyclic aromatic hydrocarbon (PAH), mercury, cadmium, and lead content; nickel release, etc.

The advanced mobile energy source to provide the sort of power we are discussing in mobile devices can be any technology that doesn't require direct connection or close proximity to a "fixed" power source. Fixed power sources include plugging the mobile device into wall outlets for charging or inductive charging of the mobile device requiring the device to be kept very close to a fixed location inductive charging source. Note that the allowable technologies could be used individually, or in combination, to achieve the usage definition as long as other requirements, such as safety and manufacturability at the target price range, can be met.

Possible advanced mobile energy technologies that could be used are the following:

1. Longer life, higher energy batteries (power storage devices)
2. Other energy storage devices such as super-capacitors, nano-flywheels, etc.
3. Energy-generating technologies such as fuel cells
4. Energy harvesting from the local environment such as solar energy, RF harvesting, or user-generated energy
5. Wireless power that doesn't require keeping the mobile device in close proximity to the power source
6. Any other means for providing energy needed by the mobile device that doesn't require attachment or proximity to a fixed location energy source

For developing innovative advanced and disruptive applications to emerge in mobile devices, a 20X or more improvement in mobile energy is essential. With a doubling every 10 years in battery energy density, this would require more than 50 years of technical development at the current development trend! This is far too long to wait for the IoT and the personal mobile technology of the near future. Other alternatives, and the approach pursued for IoT, focus on power management electronics and intermittent power for remote uses.

Work is going on in the CE industry to improve the energy density of mobile power sources, to provide wireless and other ways to charge devices and especially with more advanced energy management electronics. Of course, with consumer devices, these solutions must be accomplished with little or no addition in the bill of material component costs.

7.4 Consumer Metadata[5]

There is an urgent need for metadata to accompany and describe media data essence. For the majority of personal content, user-generated descriptors tend to be vague to the point of uselessness. Standardization in the metadata format is needed to allow

[5] Much of this material is from *A Novel Taxonomy for Consumer Metadata*, T. M. Coughlin and S. L. Linfoot, Presentation at the 2010 IEEE ICCE Conference.

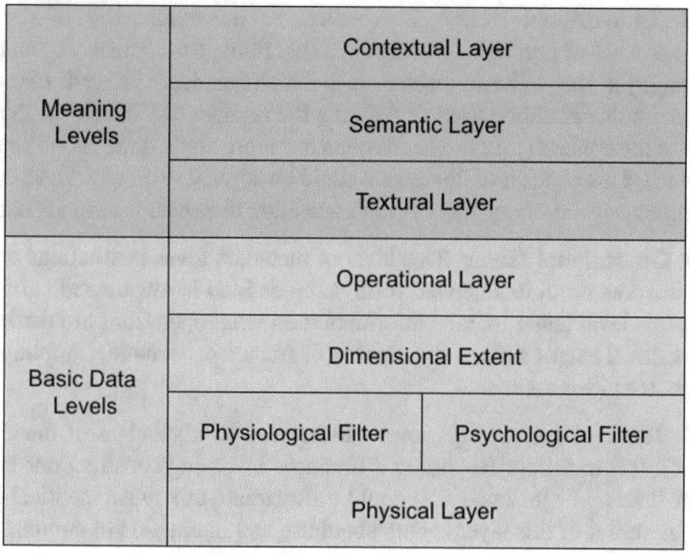

Fig. 7.3 Metadata layer model

a full and useful description of content that is interoperable between consumer devices. Manufacturers need to ensure that the metadata generated by a device is complete and understood by other products as well as allow for the creation and use of more subjective metadata.

In addition, as processing capability in the cloud and on devices increases, machine learning and other artificial intelligence algorithms can be used to generate metadata from the content itself, decreasing the dependence on human categorization.

The terminology used here treats metadata as a communication channel and has similarities to and is inspired by the OSI network model. A seven-layer metadata ontology model has the uppermost layers giving a more abstracted level of content metadata (the meaning), while the lower layers provide basic metadata (down to the physical layer of the actual input). Figure 7.3 shows a graphical representation of this metadata model. These metadata layers are described below.

Layer 1, Physical Layer (Sensory and Source Information) This is basic information about the content related to the source of the content (where and when) as well as sensory information of various sorts. The sensory information could include sound, sight, touch, or smell in some defined fashion.

Layer 2, Physiological and Psychological Filtering This metadata defines what sort of personal or experiential filtering is applied to the signal. This filtering may relate to the characteristics of the channel used to transmit (physiological) or experience (psychological) the metadata which may differ depending upon the type of data—e.g., speech or music may undergo different psychological filtering. Thus, speech can be converted to text, which can be a filtered metadata giving useful and searchable information about what was said but speech to text conversion would do little to pass on the psychological import of a piece of music.

Layer 3, Dimensional Extent This relates to the complexity of the content described as a set of orthogonal dimensions describing the content. For an image or video, this layer may indicate whether it is flat or has depth as well. Likewise, for audio content this could be used to describe the number of "voices" or the level of presence of the content (e.g., surround sound has more audio dimensions than a monaural sound). This concept of dimension could be applied to all of the senses with an interesting expansion of our ways of understanding dimensions in touch and smell.

Layer 4, Operational Layer This level of metadata gives instructions on how to recreate the content in its intended form using defined hardware and software. For instance, this level could include information on what operations are performed on the dimensional extent such as the number of frames per second, sampling rate, bit depth, etc. for video content.

Layer 5, Textural Layer This level can be seen as a subclass of the next level (semantic). It is metadata describing differences involving constructions built from the lower levels. For instance, this could differentiate otherwise identical blue and red cars. A subset of this layer is data about use and interaction of content.

Layer 6, Semantic Layer This is a concrete definition of the object or experiences in a piece of content based upon generally agreed-upon constructions—for example, "a tree"—"my friend said..." (as input from an audio byte where it is recognized that it is your friend speaking and he/she said…).

Layer 7, Contextual Layer This level refers to the description of experience of content by a sensible sentient being. Current computers cannot create true judgmental information for metadata as they cannot look at a scene and define it as "beautiful" or listen to music and define it as "melodical," analyze a smell, and refer to it as "pungent." The contextual level is by its nature subjective or personal—specific to the participant. A collection of contextual level metadata from several sensible sentient beings could be represented as providing a sort of temporary consensus on the "meaning" of that content.

The first four metadata levels are much more defined or mechanical, while the last three are increasingly individualistic or subjective. All of these levels of metadata are important in fully describing a thing or experience as perceived by people. Metadata created in these seven layers can be applied to more effectively search and use content. So, for example, a picture of a tree in a field in a bright sunny day may be defined within the metadata model as:

[Video][Visual][Still][1FPS; 640x480;RGB]{[Green][Tree][Blue][Sky][Green][Grass]}[Peaceful, Calming, Tranquil, Boring]

Another example is an interview between two people Fred and John about the weather at a party

[Audio][Background noise filtered, 50 Hz – 4KHz, speech][continuous time][44.1KHz, 16-bit, 10 seconds]{[Fred]["Nice weather today yes?"][George]["Nah, it's raining in Plymouth"][Music]}[Crowded, Noisy]

These seven metadata levels provide a way that makes it easy to search for content using all possible levels of representation. For example, it may be

necessary that a user does a search for "video of Fred talking about weather," and it will look through each layer to find the details:
QUERY : *, *, *, VISUAL, *, VIDEO, *, Fred, weather,

To make metadata general, particularly for consumers, requires automating the process of metadata generation to create metadata that is genuinely useful to ordinary people. The full value of automated metadata generation that meets the requirements spelled out in this paper will be realized when this automated metadata increases the access and use of user-generated content.

With the development in machine learning algorithms and both local and cloud-based services that can take existing content and turn it into actionable information (or metadata), we can make increasing use of existing content. The following section discusses one of the more useful current manifestation of automated content analysis (and metadata generation), voice recognition.

7.5 Voice and Image Recognition

Consumers are generally not very keen on creating manual metadata to describe content, such as a picture or video. Incorporating additional sensors in content capture equipment, such as GPS location detectors associated with a camera (like in a smart phone), can provide metadata that can be automatically captured and thus reduce some of the human error.

In addition to additional sensor-based metadata creation, advanced software using artificial intelligence technology can create metadata based upon analysis of sound or images. As pointed out previously, metadata allows easy search and use of captured content. This metadata can consist of basic information about the content or more advanced information that can put the content into context.

Voice recognition, as well as image recognition, benefits from advanced machine learning algorithms. In machine learning, a neural network can create models to make meaning from content. This is done by training these models with a variety of known content until the model becomes very accurate in its analysis. Machine learning algorithms can continue to learn from additional content as time goes on, refining its recognition capabilities.

Using these approaches, speech recognition can achieve error rates of less than 6%. With image recognition using machine learning, a recent Google demonstration achieved a 3.5% error rate. This is a lower image recognition error rate than is achieved by a normal human. Artificial intelligence software is used for image recognition as well as voice recognition. This type of voice recognition is used in many modern devices, such as Amazon's Alexa, Google Home, and Apple's Siri.

Voice recognition can be used to create text from speech. This text can be used as metadata allowing easy search for, for example, video content that includes speech. In addition to turning speech into text voice recognition can also be used to tag who the speaker is, adding additional metadata information into the content.

Likewise, image recognition can determine if there is action in a video, what sort of things are in the images (e.g., animal, woman, man, bicycle) as well as who is in the images. For more advanced implementations, these algorithms can guess at people's emotional states based upon their image and voice patterns. All this can be turned into metadata, making it easier to find and use these images.

7.6 Chapter Summary

- Modern display technology allows viewing rich media at very high resolutions. Small screens used close to a viewer's eyes can be more effective at presenting perceptible high-resolution images than larger displays at some distance from the viewer.
- To achieve the same appreciation of high resolutions at a longer distance from the viewer, larger displays are needed.
- Battery technology to power mobile devices advances at a much slower rate than applications that use that power. As a result, consumer mobile devices use very advanced power conservation that can extend battery life. Ultimately more power is needed for a smart phone to be useful throughout a day without recharging. In order to accomplish this, we need significant increases in the rate of mobile power supply scaling as well as advanced power management features.
- Metadata is a vital element in finding and using consumer content. Metadata can exist in several layers from simple descriptions of the content to more advanced information that puts the content in context and provides meaning.
- Machine learning algorithms have enabled great advances in voice and image recognition. These tools can be used to create metadata that can make content and consumer applications using this content more useful.

Chapter 8
Integration of Storage in Consumer Devices

8.1 Objectives in this Chapter

- Demonstrate the contribution of digital storage devices to the costs of consumer devices.
- Show examples of basic consumer applications that are becoming standard features in consumer devices.
- Explore trends in the development of greater intelligence in storage devices.
- Show how digital storage devices can be matched to various applications.
- Project trends leading to the incorporation of applications in digital storage devices and the economic and performance implications of this trend.

Consumer electronic devices are undergoing significant evolution. Smaller semiconductor line widths enable more functions on fewer chips. With fewer chips, consumer devices can be smaller, or they can incorporate more than one application within a single device. The content and software used in a consumer device must be stored in some sort of memory, whether that be a temporary volatile memory cache or some form of nonvolatile memory. Digital storage and memory are essential ingredients in modern consumer electronic products.

Storage devices of all sorts have larger digital storage capacity at ever lower prices, so more content can be captured or stored on a consumer product. At the same time, the cost of digital storage for many consumer devices is a significant percentage of the total bill of materials cost. Thus, cost reduction for many consumer devices will directly target digital storage. This results in pressure to lower the total cost of the storage device or alternatively it involves changing the storage devices so they play a bigger role in the end product performance, with an overall reduction in end product costs. The growth of online cloud storage provides another way to economize, where content is only accessed as it is needed.

ANSI and other standard committees create the standards that define commands understood by most digital storage products. The T13 committee defines the ATA specification used in most hard disk drives and many solid-state drives.

T.M. Coughlin, *Digital Storage in Consumer Electronics*,
https://doi.org/10.1007/978-3-319-69907-3_8

New commands are continually being added to the ATA specification allowing more complex commands that enable features such as media streaming in consumer electronic products.

Standards are also being developed for solid-state storage devices in order to perform various consumer functions as well, particularly NVMe standards that run on the PCIe interface. As consumer functions have become standardized, features such as GPS, radios, and DVRs can be implemented as firmware in a single chip or a small number of chips with appropriate supporting external hardware and software applications.

Digital storage hybrid products that combine more than one digital storage technology in a device are used in some applications, such as hybrid hard drives with flash memory incorporated into the hard disk drive or on the motherboard. These flash caches speed boot times and reduce power use by aggregating writes onto the drive or provide fast access to frequently read data. Such flash cache injections into storage architectures are examples of how combining different elements from the consumer electronics digital storage hierarchy can result in a product with features neither storage device could have alone.

These developments create opportunities to use digital storage in new and interesting ways in consumer devices. Clever design of storage devices in consumer products provides greater functionality and performance, lower overall end product cost, higher reliability, and smaller product footprint. This chapter explores the development of storage for consumer products today as well as looking where storage integration concepts could develop in the future.

8.1.1 Storage Costs in Consumer Product Design

The following tables are representative hardware component *bill of materials* costs as a percentage of the total hardware cost for several common consumer electronic devices. These devices include digital storage in their design in order to support the availability of content that makes these products useful. These examples are shown to give an appreciation of the percentage of the cost of the device that is due to digital storage and memory. As we shall see, for many consumer products, digital storage represents a significant part of the *total product cost*. This will have implications on future design of digital consumer electronic devices as shown later in the chapter.

Table 8.1 gives a typical percentage breakdown of the percentage of total product cost for each of the key components (a bill of materials or BOM cost) in a representative set-top box with DVR capability. Peripheral components such as cables and external storage are not included since these costs can vary and often are aftermarket products. As can be seen in this table, digital storage is a significant percentage of the total system cost for a digital video recorder and a larger percentage than any other single component.

Table 8.1 Set-top box with DVR component cost percentage of total system cost (Represents composite data from various sources)

Component	Percentage of total system cost
Motherboard	40.4
Hard drive (500 GB)	41.9
802.11n card	5.8
Power supply	2.2
Other	9.7

Table 8.2 Portable music player component cost percentage of total system cost (Represents composite data from various sources)

Component	Percentage of total system cost
Display	3–5
Internal memory (digital storage)	65–75
Battery	2.0–3.0
Power management electronics	1.5–3.5
DSP electronics	4–6
Interface electronics (such as USB)	0.5–1.5
Audio CODEC	1.5–3
Housing and other mechanical parts	3–6

In an automobile navigation and entertainment system, the proportion of the BOM costs represented by the hard disk drive and various solid-state memories is 6–15%. Peripheral components are not included since these costs can vary and often are aftermarket products.

Table 8.2 gives a typical percentage breakdown of total product cost for each of the key components (the bill of materials or BOM cost) for a mobile music (MP3) player with minimal display capability. Peripheral components such as headphones, speakers, microphone, removable memory, and protective case are not included since these costs can vary and often are aftermarket peripheral components. As can be seen in this table, digital storage provides the majority of the total product cost for a mobile music player.

Table 8.3 gives a representative percentage breakdown of the total product cost for each of the key components (the bill of materials or BOM cost) in a representative smart phone (with 128 GB NAND capacity). While not as large a percentage of the total BOM cost as the other products, digital storage represents the single largest percentage cost item in the BOM.

Table 8.4 gives a representative percentage breakdown of the total product cost for the key components (the bill of materials or BOM cost) in a representative tablet computer with 128 GB storage capacity as well as Wi-Fi and 4G connectivity.

As can be seen from these tables, digital storage provides a significant portion of the total cost of many modern consumer devices. In many cases, it is the single

Table 8.3 Multimedia mobile phone hardware component cost percentage (IHS teardown of Google Pixel phone, October 2016, Engineering.com)

Component	Percentage of total system cost
Internal memory (NAND flash and DRAM)	27.6
Video and audio player electronics	13.0
Wireless and telephone electronics	6.4
Other PCB components	4.6
Camera module	4.6
Display	15.1
Battery and power management	3.9
Electromechanical and mechanical parts	15.4

Table 8.4 Multimedia mobile phone hardware component cost percentage (IHS teardown of Apple iPad Air 2, October 2014, cnet.com)

Component	Percentage of total system cost
Internal memory (NAND flash and DRAM)	20.7
User interface and sensors	6.3
Wireless and telephone electronics	10.8
Processors	6.5
Camera module	3.2
Display and touch screen	33.1
Battery and power management	6.7
Electromechanical and mechanical parts	12.7

most expensive component in the device. Digital storage capacity is likely to continue growing to meet the need for higher resolution content. Although many users are now moving their content into the cloud, there are some users who need more storage capacity, and for them, the total cost may be dominated by the cost of storage.

8.2 Development of Common Consumer Functions

Mature applications become standardized, and with the increasing chip complexity, the number of electronics chips required to implement those applications decreases. This section explores some important consumer functions that are becoming ubiquitous on consumer electronic devices and the role and type of digital storage used in these applications.

The emerging standard functions that will be examined here are:

- Digital video recorders
- Digital cameras
- GPS location services
- Wired and wireless network connectivity

8.2.1 DVR as a Standard Consumer Function

Not too long after the appearance of the original DVRs in 1999, designers began to look at combining DVRs with other entertainment systems and emulating the record and playback functionality of DVRs for non-video applications. DVR functionality began to be integrated in some DVD player/recorder devices to provide a single box DVD player and recorder as well as digital video recorder. DVR functionality has also been directly built into some TVs.

In general, the content recording functionality is becoming one of those iconoclastic functions that are now incorporated into many products. TV tuners and appropriate software such as Microsoft's Home Media Center have been used to make desktop and laptop PCs record and playback television programs. There are also digital radio audio recorders offering an audio equivalent of digital video recorders. Today set-top box and other products that integrate DVR functionality into another product are much more common than stand-alone DVRs.

As a function such as digital recording becomes more widespread, the basic electronics to support the function are incorporated into general purpose electronic ICs. The function can then be enabled by appropriate firmware (software that runs inside a device). This bundling of maturing functions into chip sets enables such applications as DVR or audio recording on space-restrained devices such as mobile phones with TV reception.

Also, the empty space in static home DVR devices can be removed allowing smaller set-top boxes or DVRs. This reduction of space makes it much easier to include DVR functionality into more consumer devices. Note that the height of the hard disk drive may restrict the reduction of thickness of the host device, but other dimensions can be reduced as needed.

8.2.2 Cameras as a Standard Consumer Function

Digital still and video camera capabilities are ubiquitous in consumer devices such as cell phones, tablet computers, and computer laptops. Unless zoom and high magnification are required, the optics to create a camera are now easy to implement in a restricted space such as in a cell phone. Likewise, the electronics to make a camera have been minimized and incorporated into standard SoC electronic chips, adding minimal cost to enable these functions.

Today almost everyone carries a video and still camera with them in their cell phone, and many people use these applications regularly. As these cameras have begun to show up on cell phones and other products, people are using them to capture photographs where ever they go and share them. The increasing presence of digital still and video imaging capability in more and more hands is one of the drivers of the social networking and content sharing that is so popular with people of all ages.

Cameras are a part of many consumer devices and are used for many purposes. This includes content capture as well as image and facial recognition. Connected devices with cameras will be used for new social networking and content capture applications and for things we haven't even thought of yet. This is a technology that is a standard consumer product function.

The growth in the number of cameras that people carry with them opens up some interesting possibilities for continuously recording your life, creating a *life log* or for using multiple images from cameras to create different perspectives of events and activities or even to recreate a 3D image of these events or activities with appropriate software.

8.2.3 GPS Location Services as a Standard Consumer Function

GPS has enabled location-based services in many products such as cell phones and still and video cameras. As a consequence, the cost of GPS capability is minimal with the electronic functions built into standard SoCs and may not even show up as a discrete BOM component. As a result, the cost of implementing GPS is only the cost of an additional built-in antenna. Consequently, location services are part of many consumer electronic products.

Location-based services allow ways for people to connect with each other, electronic navigation, and new ways to sell services to consumers.

8.2.4 Network Connectivity as a Standard Consumer Function

Most mobile and static consumer devices include wired or more likely wireless access to local networks, usually with Wi-Fi. Many static consumer applications such as a media server are built around basic network connectivity. For mobile devices, wireless connectivity using Wi-Fi and/or Bluetooth technology is the natural way to provide local or Internet network connectivity. In cell phones and many other products, an additional path for connectivity is through wireless phone networks such as 4G and 5G.

Network connectivity allows a consumer device to exchange content with other devices on the local network or over the Internet. It allows synchronization and backup of content on these devices as well as streaming content from one source to another. It also allows the development of what is called the Internet of Things (IoT), where smart connected devices interact with each other and people to provide services and entertainment. We will discuss the role of digital storage in the IoT in Chap. 9.

Practical digital rights rules that are agreed to by most users and content owners and providers as well as privacy protection will be critical to the full realization of the potential of network connectivity and content delivery.

8.3 Intelligence of Digital Storage in Consumer Electronics

The capabilities built into storage devices such as hard disk drives and flash memory have been increasing. This is because semiconductor line widths have declined, allowing more electronic functions on a given chip. As a consequence, more functions are built into and enabled by the firmware in a digital storage device. As an example, the T13 committee ATA specification for hard disk drives includes commands that enable multiple video and other richer streaming content so that recording, sharing, and playing of content can go on simultaneously.

8.3.1 Security Providers

Data storage device companies and those that design the products that they are incorporated into have been working together to increase the security of storage devices in mobile products (such as laptop computers). The *Trusted Computing Group*[1] is one of the key standards organizations working to make such products more secure. The influential storage and host device companies that are members of this standards effort have created and published a Trusted Storage Specification. One application of this provides encryption of the user data where all aspects of the data encryption, including key management, are done on the storage device (either a hard disk drive or a solid-state drive).

This standard covers any storage device that incorporates security and trust services running within the storage device. The first products that were covered by this specification were HDDs, but their use in SSDs is greater because crypto-erase by erasing the embedded encryption key is one of the few ways to remove most of the data on an SSD.

With this hardware encryption method, the encryption key never leaves the storage device. The encryption key and the encryption firmware are stored on a nonuser accessible area on the storage device. Unless the user accesses the data using a password that enables the encryption key, they cannot get to the data. If the encrypted key stored in the nonuser accessible area of the drive is intentionally written over by the user, the data on the drive is not accessible.

The data encryption function on the hard disk drive is carried out by an application running in the storage device called a *security provider* (SP). Security providers can be used for various security and trust functions or commands besides encryption

[1] www.trustedcomputinggroup.org/home

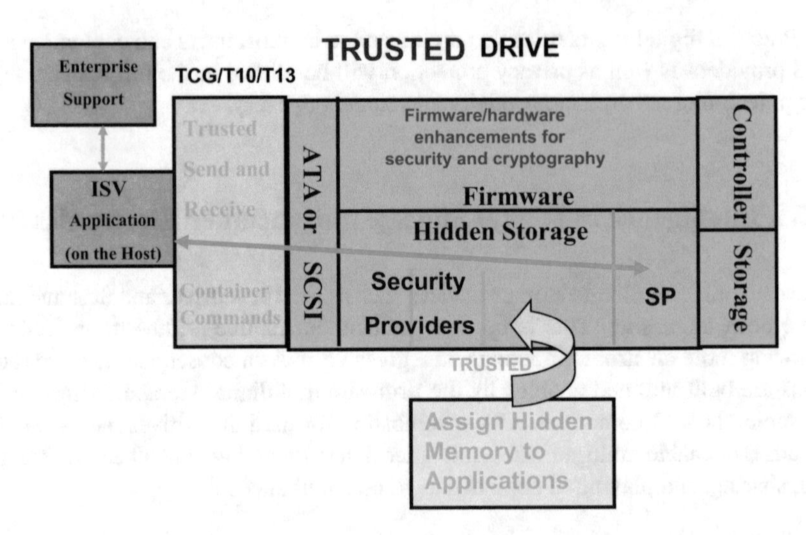

Fig. 8.1 Architecture overview of trusted storage showing security providers (Michael Willett of Seagate presentation at 2007 Data Encryption Forum)

services and become a root of trust associated with the storage device. SPs are located on partitioned regions of nonuser accessible storage on the storage device. The hidden memory in one of these hidden partitions can also be assigned to an application or more likely a basic application command since the size of the SPs are not very large (20–30 k bytes).

Use of such application partitions combined with integration of application hardware functions on a storage device could be important elements in the development of more integrated consumer devices. If desired, such security providers would be a natural location for various applications that can run before general system boot. Implementation of this architecture is achieved through interface specifications such as T13 for ATA and T10 for SCSI as well as the Trusted Computing Group. Figure 8.1 shows the general architecture overview of the trusted storage security provider.

8.3.2 Object-Based Storage

A key mode to increasing the intelligence and capability of consumer electronic devices is the use of metadata. Metadata is data about data that summarizes and makes more useful the original data. Traditional metadata is file headers and other information about dates of creation and modification of a file, the software that was used to create it, and who the person is that made it. More sophisticated metadata, such as that used in the creation of movies and videos, includes information on location, who is in the images, timing information, and other data that will be useful in editing the content.

In a digital storage device, metadata plays a similar important role in telling the hard drive what sectors of data blocks must be combined to create an actual data file. At the present time, although this metadata may be stored on the storage device, the actual reconstruction of content and the use of the metadata to organize and use the content are done in the microprocessor of a host device. The software that is used to recreate the data file from the raw blocks of data using the file metadata is called a *file system*.

If a storage device included its own version of a file system, then the storage device would be able to access the data at least in part without the assistance of a host file system. If such a capability was built into a digital storage device, that device could do many things that currently are done on the connected host device. This would include functions such as searching for a file. If a storage device could automatically *index* (create searchable metadata) the files that it contains, then the storage device could off-load, augment, or potentially even replace host system functions such as searching for files and content.

If indexing software can be created that can sort and categorize still and video images and even sort by people or locations from a user's own database of information and experience, then the potential capability of finding and using content in the storage device itself could be awesome.

A storage device that includes such a file system is called an *object-based storage* device, and the files on the storage device are referred to as the objects. The SCSI specification has included support for object-based storage devices for several years, but the ATA specification used in many consumer storage devices such as hard disk drives has not included object-based storage support. Figure 8.2 compares a traditional host-based file system to an object-based storage device. Note that for the object-based storage case, the file system is divided between the host and the storage device.

By including at least some of the file system in the storage device, the host system can be freed to perform other functions providing faster host functionality and potentially much faster performance. Imagine if there are several storage devices in a system such as a hard disk drive or SSD array and if when a search request was made for files with some particular data the disk drives could independently look for the data—the potential speed up in the search could be significant.

Object storage devices are still not common in consumer applications, although they are supported in various operating systems. Object-based storage at the system level is now common in enterprise storage, particularly for large storage systems such as active archives.

Electronic integration (e.g., more powerful microprocessors on the storage device electronics) enables object-based storage on individual storage devices. Seagate's Kinetic drives with Ethernet interfaces are one example of an object-based storage device where the drives manage their own content. Western Digital's WDLabs has introduced similar Ethernet interface drives and demonstrated them running a Ceph cluster. The same approach could be made with solid-state drives. These object-based storage devices are currently intended for enterprise storage applications.

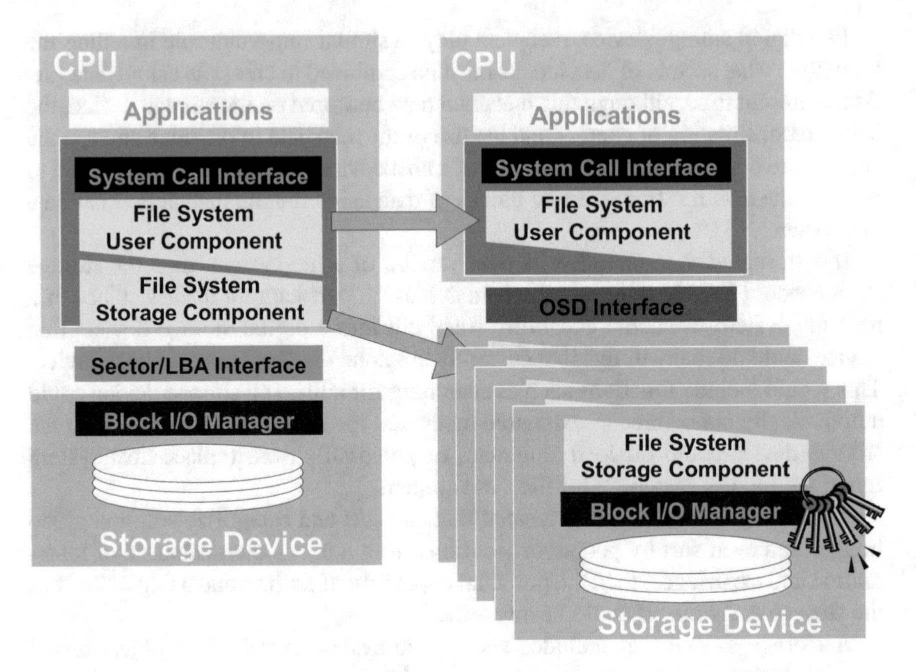

Fig. 8.2 Comparison of traditional file system storage to object-based storage (Michael Willet of Seagate presentation at 2007 Data Encryption Forum)

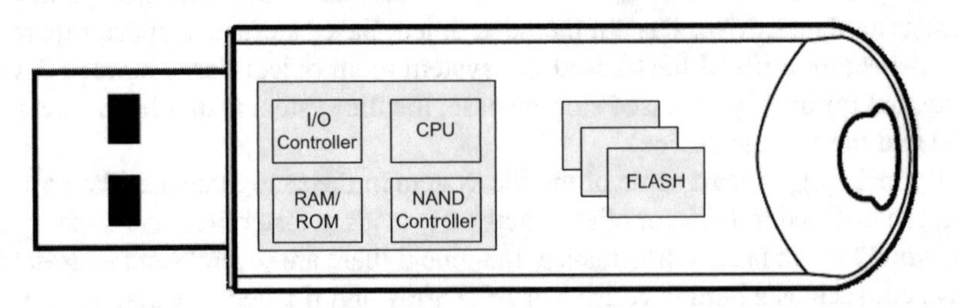

Fig. 8.3 Schematic of a USB drive showing internal components that make this a small computer

There are no object-based consumer storage devices at this time, but perhaps in the future this will change with the development of IoT gateway products in homes that might benefit from object-based storage where the storage device can manage and find its own content.

8.3.3 USB-Run Software Applications

Flash memory devices are being created with greater intelligence and functional capability. In essence, a USB device is like a small portable computer as shown in Fig. 8.3. Because a USB drive has its own CPU and a file system, it is capable of performing applications.

In 2003 the *USB Flash Drive Alliance* (UFDA) was formed; it was a consortium of leading USB flash drive manufacturers, supporting USB "smart" drives, that allowed users to run active programs from USB flash drives. USB smart drives provided users similar functionality to desktop applications run from a flash drive. Lighter-weight programs such as games, instant messaging and photo editors, browsers with personal setting and bookmarks, as well as encryption were run from such smart USB flash drives.

UFDA members included Lexar; PNY Technologies, Inc.; Samsung Semiconductor Inc.; Buffalo; Crucial Technology; Infineon Technologies; International Microsystems Inc.; Kingston Technology; and Peripheral Enhancement.

In 2006 SanDisk (and M-Systems, since acquired by SanDisk) developed their own competing application platform for USB-based devices called *U3*. Microsoft and SanDisk created a successor to U3, called SmartKey. SanDisk withdrew support for U3 starting in 2009.

Flash memory devices can contain data that is private or proprietary. Thus, as with hard disk drive storage, flash memory devices with built-in technologies to protect the contents from non-authorized access are in the market. There are many USB storage devices with built-in encryption.

USB drive-based applications, particularly booting a computer and running it from a USB drive, never took off, probably because USB drives with powerful enough processors were too expensive for most consumers. However, software on a USB drive plugged into a computer is commonly used to run software that the user doesn't want to install into the host computer or to keep data off the host computer.

8.4 Matching Storage to Different Applications

For various reasons, different types of digital storage devices are better matched for different sorts of applications. Table 8.5 compares hard disk drives and flash memory storage capacities available for various enterprise, computer, and consumer applications.

The choice of a digital storage device for an application is the result of an engineering and economic trade-off. As described in Chap. 1, as the technical and economic requirements for an application become clear, there are clear demarcations between the appropriateness for different types of storage devices. Thus, there develops a storage hierarchy for consumer electronic applications, having some similarity to memory and storage hierarchies in computers and enterprise applications.

This hierarchy is only a guide since there can be a great value in achieving some of advantages of various storage devices by combining them into some sort of hybrid storage device containing more than one storage technology. The use of flash memory cache on a hard disk drive is an example of such hybrid storage technology (it should be noted that an alternative approach sometimes supported by Intel and others is to put the write cache on the host motherboard).

Table 8.5 Storage device and application requirements (hard disk drives and flash memory)

Application	2017 storage capacity	2021 storage capacity	Capacity need potential	Flash application	HDD application
Enterprise storage	HDD: up to 2.4 TB (SCSI/FC), up to 14 TB (SATA) SSD: up to 16 TB (NVMe)	HDD: up to 4.2 TB (SCSI/FC), up to 21 TB (SATA) SSD: up to 60 TB (NVMe)	Unlimited driven by on-premise and hyperscale cloud application and compliance	Yes, eventually as primary storage and certainly for performance applications	Yes for a while for cost-effective performance and later as low-cost capacity storage
Desktop computer	HDD: up to 8 TB SSD: up to 2 TB	HDD: up to 14 TB SSD: up to 4 TB	Declining market except for professional high-performance workstations	SSDs gradually displacing or supplementing HDD storage	Desktops have room for tiered SSD and HDD storage
Notebook computer	HDD: up to 2 TB SSD: up to 1 TB	HDD: up to 5 TB SSD: up to 4 TB	Stable market, local versus cloud storage drives internal storage needs	SSDs gradually displacing HDD storage except for external storage (e.g., backup)	HDDs still dominant but SSDs gaining as flash prices decline
DVR/ set-top box	HDD: up to 2 TB	HDD: up to 4 TB SSD: up to 2 TB	Network DVR could displace DVR; capacity demand could be large if external storage enabled	Could be some opportunities for flash memory	HDDs dominate because of costs of storage. If external storage enabled, this could be HDD volume driver
Music player	Flash: up to 128 GB	Flash: up to 128 GB	For just music capacity growth is limited	Yes, dominant storage technology	No longer used in music players
Tablet	Flash: up to 512 GB	Flash: up to 2 TB	Video and other rich content is big capacity driver	Yes, dominant storage technology	Never found a niche
Smart phone	Flash: up to 512 GB	Flash: up to 2 TB	Video and other rich content is big capacity driver	Yes, dominant storage technology	Never found a niche
Digital camcorder	Flash: up to 512 GB	Flash: up to 2 TB	Video is big storage driver	Yes, dominant storage technology	HDDs have been displaced
Digital still camera	Flash: up to 256 GB	Flash: up to 1 TB	These do video too	Yes, dominant storage technology	HDDs have been displaced
Personal life recorder			This could use TBs of storage	Probably use flash for mobile use	

8.5 The Convergence of Electronics: When the Storage Becomes the Device or Was It the Other Way Around?

The integration of device functions into smaller and smaller electronic packages has been enabled by shrinking of semiconductor line widths as well as improvements in grounding and signal isolation in complex chip designs. Integration of hard disk drive electronics has allowed creating circuit boards that do not occupy the entire back of the hard disk drive like older disk drive generations. With higher unit volumes, expected developments in electronic circuit densities, and clever design, it is reasonable to expect that consumer application electronics as well as storage device electronics could fit on a circuit board no larger than the dimensions of the storage device. Likewise, with denser electronics, flash-based devices could also be built with tighter integration into applications as well.

As consumer applications mature, they become more standardized. Standardized functions can be implemented on a chip with the functions accessible through firmware. As more and more consumer applications can be built into a few chips, it is a natural step to look at how the digital storage can be most effectively integrated to reduce the complexity of the devices, improve their performance, and lower their cost.

One example of how this could be enabled for a mobile personal media player is shown in Fig. 8.4. In this figure, the consumer application electronics are incorporated in the electronics of the storage device itself, in this case the circuit board of a hard disk drive.

For applications where the digital storage is a significant fraction of the total cost of the end device, such an integration of functions would make these devices cheaper to manufacture, since the functions are built into the storage device during its manufacture. In addition, integrating features enable the CE functions to be tested during the storage device burn-in. By not having to make a separate circuit board for the CE device, the size of the end product could be minimized, and the cost of components and assembly of the CE device circuit board could be eliminated.

Since lead lengths on a common circuit board would be minimized, performance should be better in an integrated device. If designed properly, programming unique functions and combinations of functions for the end product would be easier to do than in normal product integration (involving mostly firmware changes). These advantages could make such approaches, combined with object-based storage devices, commonplace within the next few years.

Although we used a hard disk drive as the example, flash memory electronics or other storage product electronics could use a similar approach to lower overall end product costs and providing better performance and product design.

There are many advantages of tighter integration of the storage device and application electronics:

- Reduction of the bill of materials (BOM) cost through reduction of the number of host circuit boards and chips.
- Reduction of manufacturing steps for the host device through assembly in a single circuit board or controller device.
- Assembly of the host device in highly efficient storage production plants (such as hard disk drive or solid-state drive manufacturing plants).

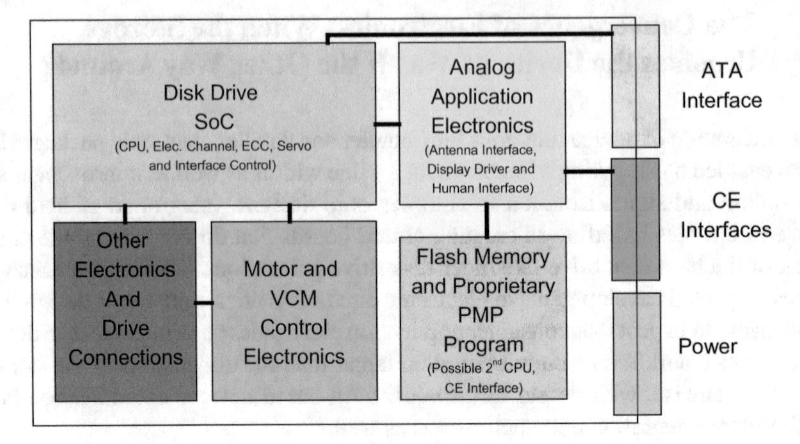

Fig. 8.4 Example of a personal media player (PMP) implemented on a storage device

- Testing of the consumer product applications as part of the storage device production testing, thus eliminating additional application testing.
- Tight integration of the storage device with the host application should allow considerable improvement in host device performance since the interfaces between the application and the storage are in close proximity, and with proper standards, the rules for creating optimized interfaces will be well known.
- With shorter leads and higher levels of integration, it should be easier to control application power usage and thus prolong the life of the mobile power source.
- An integrated application in the storage device could allow ancillary capabilities building off of existing hard disk drive technology such as application SMART (Self-Monitoring and Reporting Technology) that can log application performance in a storage device log file and flag potential failures of the application modules.

8.6 Road Maps for CE Application Integration in Storage Devices

Let us look at two cases of storage device/application integration: single storage device integration with applications and integration where there is a collection of two or more storage devices that are integrated together. The following sections describe these approaches.

8.6.1 Single Storage Device Application Integration

Integration of consumer applications into storage devices will have tremendous economic pull and provide new levels of design efficiency and improved performance. It will probably take a few steps to reach the ultimate levels of integration. Thus, we

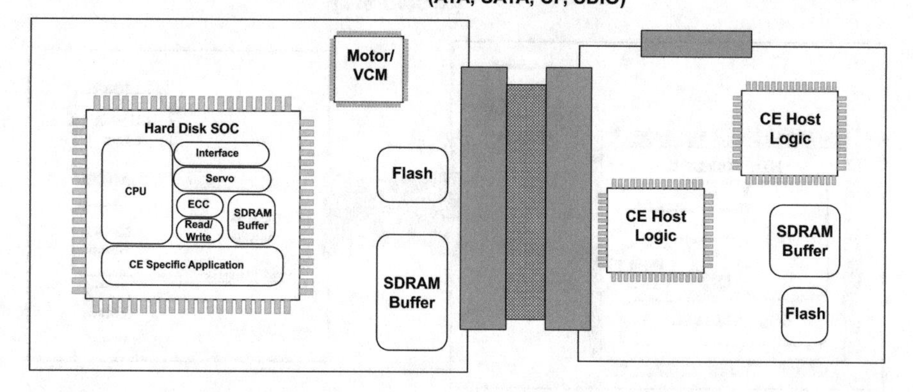

Fig. 8.5 Phase 1: Today's common approach to integration of storage devices in consumer electronics

believe that there will probably be phases of integration as described below.[2] The example given below is for hard disk drives, but this idea could equally apply to flash memory and other storage devices.

Note that Phase 1 (Fig. 8.5) is the current status for many consumer electronic products that include digital storage. Phase 2 (Fig. 8.6) is an intermediate step in storage integration where a storage device SoC, designed for CE applications, includes some capabilities for power control, caching of data, and other functions that assist in the function of the consumer electronic products that they are integrated in. This has been limited adoption of this phase.

Phase 3 (a, b, and c) (Figs. 8.7, 8.8, and 8.9) sees the rise of true integration of the application electronics on the storage device circuit board. We present three variations (a, b, and c) of this integration of the CE applications on a storage device circuit board depending upon the cost/price sensitivity of the consumer applications vs. application performance and ease of development.

Phase 1

The consumer electronic manufacturer/ODM purchases standard storage devices from a HDD or SSD supplier and integrates the storage device as a component into the final system. This is the current state of the industry as storage devices are leveraged for the consumer market. Each consumer electronic product platform group must develop/optimize drivers, file system, and power management features associated with the storage device. The cost structure of this architecture leads to a higher bill of materials cost than follow-on stages.

[2] Pat Hanlon, Coughlin Associates Report, 2004.

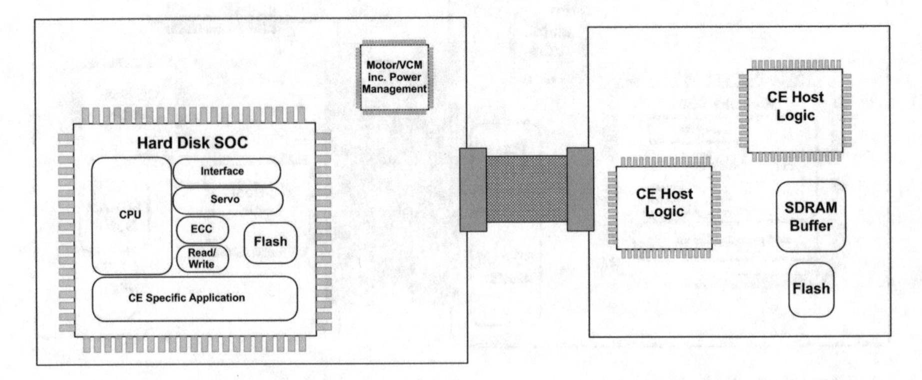

Fig. 8.6 Phase 2: Hard disk drive system-on-chip (SoC) on drive circuit board designed for CE applications

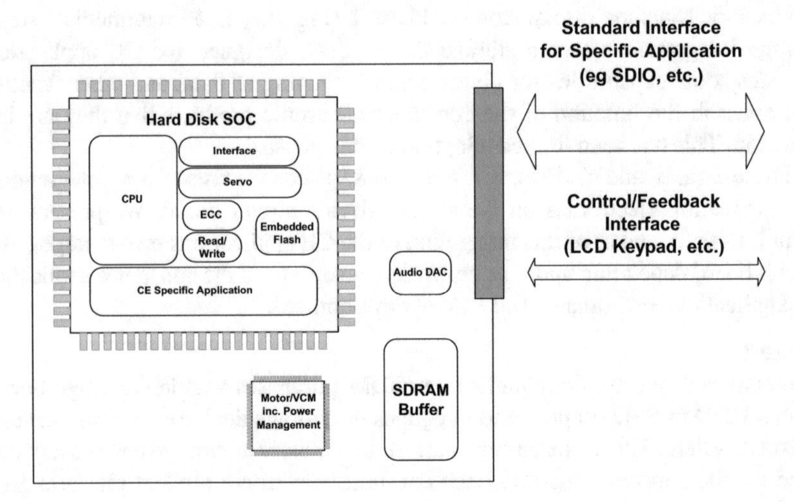

Fig. 8.7 Phase 3: True integration of consumer electronic applications on storage device circuit board

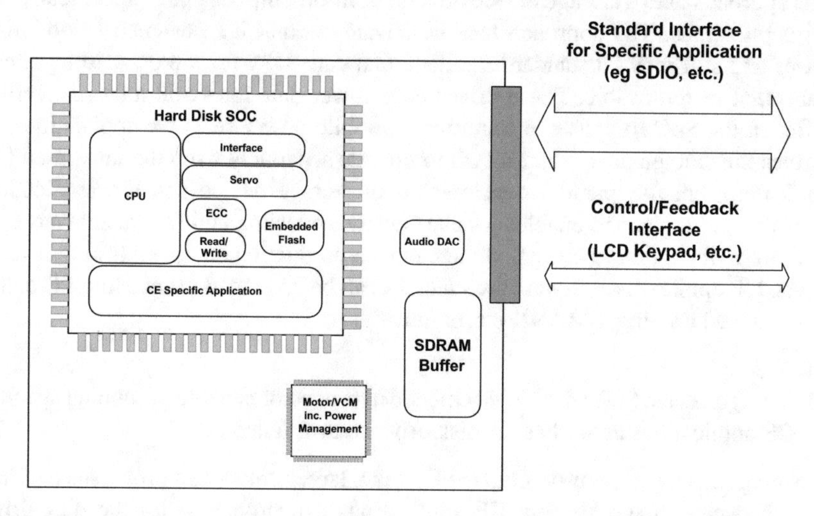

Fig. 8.8 Phase 3: True integration of consumer electronic applications on storage device circuit board

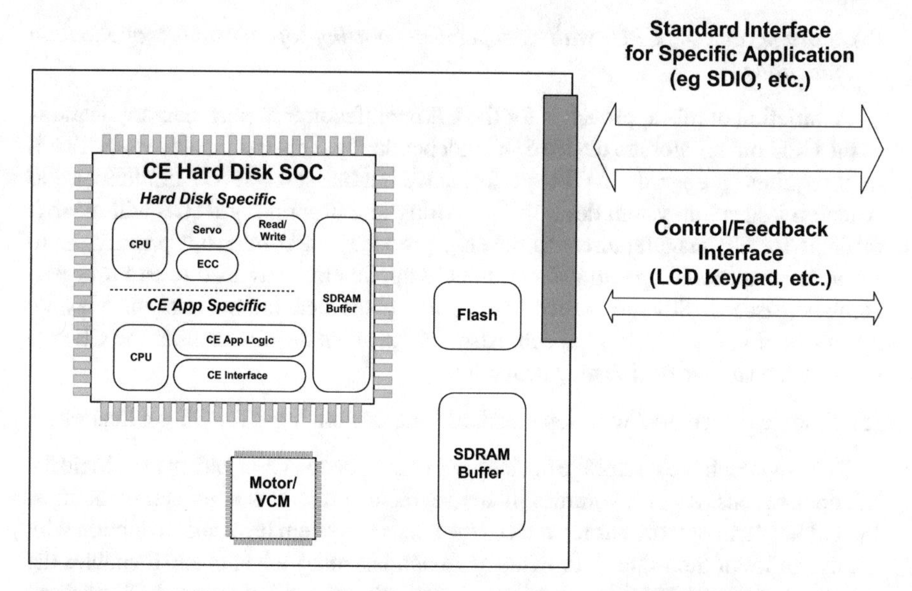

Fig. 8.9 Phase 3: True integration of consumer electronic applications on storage device circuit board

Phase 2

This approach uses a storage device SoC (system-on-chip) designed specifically for CE applications. The approach has the advantage that it's designed to optimize power and still retain a standard interface that can allow for the drive to be either embedded or removable. These drives have lower data rates and a small internal buffer in the SoC to conserve footprint in mobile power-sensitive applications. It enables the storage device and RTOS to operate separately from the integrated CE application, enables parallel development of storage device and CE application firmware and logic, and enables a clear division in testing and failure analysis that minimizes disruption to current storage device production back-end testing investments. CE application software architecture can be leveraged off existing PC code, with reduced caching, SMART, etc. commands.

Phase 3

(a) Storage device SoC with single embedded processor capable of running specific CE applications as well as the disk drive control functions

Storage device system-on-chip (SoC) has a larger embedded processor that has enough cycles to run specific CE applications synchronous with the disk drive embedded system. This approach has significant merit for more mature and standards-based applications such as MP3 recording and playback. Semiconductor providers must provide integration for this approach to be successful. The CE application portion will be pretested at the chip level so that standard storage device production and testing procedures can be utilized.

(b) *Storage device CPU with independent porting of consumer application functions*

A variation of this approach is for the CE manufacturer to port their applications to the CPU on this storage device SOC independently. This has the advantage that it only requires an upgrade in CPU performance but has some severe limitations and potential issues for system developers. Writing to a proprietary RTOS will make it difficult for CE manufacturers to develop products. CPU interrupt priority set to favor internal device operations over the CE application may lead to end user performance issues. Storage device testing processes and failure analysis may be adversely affected by this approach. Also, independent development of the CE system and the storage device may be harder.

(c) Storage device SoC with separate CPU and custom logic for CE applications

This approach uses a storage device SoC with a second CPU and custom logic for CE applications. As unit volumes in key segments grow, this approach is the most likely long-term outcome as it balances the needs of system level cost reduction with the dynamics of the value chain. This approach has many advantages. It enables the storage device and its RTOS to operate separately from the integrated CE application, enables parallel development of storage device and CE application firmware and logic, and enables a clear division in testing and failure analysis that minimizes disruption to current hard drive production back-end testing investments.

It will take some time for the evolution of these various phases of storage integration in single storage device-enabled consumer electronic devices. The phase of complete integration of the applications on the storage device will take some time to dominate the market, and this trend will be greatly assisted by a coordinated effort for open standards that make the integration easy and attractive to the consumer electronics companies while at the same time providing for new markets for digital storage devices and ways in which the companies making these digital storage devices can add additional value to their customers and perhaps enjoy higher margins as a result.

Although the storage devices used in this illustration of the path of integration into consumer electronic devices are hard disk drives, the same trends could be expected of other storage devices such as SSDs and even optical disk storage devices, where use of these storage media is appropriate.

With electronic integration, it will eventually be possible to incorporate significant digital and analog processing on the same chip as the flash memory or at least in chips that are part of the same chip stack. Such products could provide application integration such as that discussed above.

8.6.2 Multiple Storage Device Application Integration

For multiple storage device-enabled consumer electronic products such as home media centers and home network storage products, it may be less costly to take the other extreme in CE storage integration. In this case, if the control and other electronics of the multiple storage device products are integrated on the host product electronics, this results in a lower total product cost. Thus, the controller costs of the individual storage devices can be aggregated on the host device circuit board, saving some cost on the total hardware.

The cost savings likely will be somewhat less than for the integration on the storage device for single storage device-enabled electronic integration. The author estimates that total BOM savings could be 10–15%. This may not be worth the effort to develop and implement this approach.

8.6.3 Chapter Summary

- For many consumer applications, digital storage is a critical component, and the digital storage devices are a significant percentage of the total manufacturing cost of the consumer device.
- Applications that are becoming standard features in consumer devices include digital video recorders (DVR), digital still and video cameras, GPS location services, and network connectivity.

- The common storage devices used in consumer devices are becoming more intelligent.
- New intelligent features in hard disk drives include security providers (SP) in the Trusted Computer Group specifications, distributed files systems, and object-based storage.
- Intelligent features in flash memory include flash file systems and USB drive applications.
- The matching of different storage technology to various applications is shown for several computer and consumer applications including hybrid storage devices.
- An example is given of how a personal media player can be integrated on a storage device circuit board.
- A detailed road map of the development of application integration on single storage device products shows how integration of application and storage device electronics could occur.
- Some comments on sharing resources for multiple storage device products to reduce costs are given, and the limitations of these approaches are explored.

Chapter 9
Home Network Storage, the Cloud and the Internet of Things

9.1 Objectives in this Chapter

- Discuss trends in home networking including performance and digital storage capacity requirements.
- Present various implementation options for home network storage with comparisons and analysis.
- Evaluate the major drivers of networking in the home including media sharing and home reference data backup.
- Discuss the impact of consumer IoT and artificial intelligence on home networking and digital storage requirements.
- Show why human-related content is larger than any other type of content and why this trend will increase in the future.
- Show typical network storage devices, and discuss the design of network storage devices for the home.
- Project the growth of direct-attached and network storage devices in the home.
- Introduce the operation of home gateways and storage.
- Look at the options for providing digital rights protection for commercial and quasi-commercial content as well as privacy for non-shared personal content.

Digital storage in the home was once only found in home computers. Home computers might be networked together in order to share a broadband connection and in some cases to allow for network backup in the home. As more consumer devices are connected and use digital storage, there will be a greater incentive for home networking to support sharing of media between these various digital CE devices and as a buffer between local connected devices and the cloud. Although there are clear requirements for digital storage and networked storage in the home, the requirements for home network storage will be very different than those for

T.M. Coughlin, *Digital Storage in Consumer Electronics*,
https://doi.org/10.1007/978-3-319-69907-3_9

network storage for businesses. Home network storage must work with little home owner intervention and must be very reliable in order to be attractive.

A network that connects multiple storage devices can also be the basis of a system for organizing and managing the content on the devices that are connected by the network. *Automatic metadata generation* to organize content for efficient searching can be important to let devices on the network know what is on various storage devices. This data can be used to find material, to efficiently back up content on the devices without making multiple copies, to make sure that the right content is on various devices, and generally to make all of the storage devices on the network part of a unified home storage pool. This metadata generation may be done in the home, or increasingly, using artificial intelligence running on powerful servers in big data centers.

This chapter explores the methods for the design of home storage networks and devices that connect to these networks. It also looks at future developments for home network storage including how storage networks can be used to simplify the management of consumer devices and overall storage content in and around the home.

9.2 What Drives Home Networking Trends?

There are several factors that lie behind the growth of home networking. Many homes have more than one PC. Also, many homes now have wired or more likely wireless connections to local networks and the Internet. Broadband connections (typically cable and DSL) are common for most homes enabling content delivery services, such as streaming content.

In addition to sharing the Internet, network hardware, and entertainment, there is another factor that may drive home networking and also home network storage. Consumers are generating vast and growing quantities of personal digital content such as digital photos, home digital movies, documents, letters, emails, etc. In many cases, these currently exist on single computer hard disk drives. If that hard disk drives were to fail, this irreplaceable personal "reference" content would be lost forever.

Backing up the data on the hard disk drives in computers and even in home media devices is very important. This can be done on external hard disk drive boxes attached to single PCs or into a cloud storage system. Another option is to back up this home reference content on a network storage device.

9.3 Networking Options in the Home

There are several options for networking of computers and entertainment devices in the home. In Table 9.1, we summarize some characteristics of these network technologies.

Table 9.1 Characteristics of home networking alternatives (2017 Coughlin Associates)

Network technology	Reliability	Data rate	Comments
T-100 Ethernet	Excellent	~10 MB/s	Common network technology
1 gigabit Ethernet	Excellent	~100 MB/s	Used in newer installations
10 gigabit Ethernet	Excellent	~1000 MB/s	Faster but more expensive
Wireless [802.11x]	Variable	< 6.8 MB/s (a and g) <125 MB/s (ac)	Unlicensed band, interference
HomePlug	Good	<25 MB/s (AV) ~62 MB/s (AVw/ proprietary extensions)	Network connection when plug into power. AV is audio-video version
HomeGrid (merged with HomePNA)	Excellent	<125 MB/s (extensions)	Network connection when plug into telephone, coax, or power line
Coax	Excellent	~300 MB/s	MoCA 2.5 standard data rate

Table 9.2 Data rates for home media streams (Coughlin Associates)

Home media stream	Average data rate (MB/s)	Comments
MP3 audio	0.02 MB/s	Typical MP3 12:1 compression
DVD audio	1.2 MB/s	Very high fidelity digital audio
DVD video	1.25 MB/s	MPEG-2 video
Blu-ray video	5.0 MB/s	Blue-ray disk max. Data rate
SDTV	0.12–0.28 MB/s	Commonly quoted data rate
HDTV	0.28–0.56 MB/s	Commonly quoted data rate
UHDTV 4 K video	1.3–3.1 MB/s	Commonly quoted data rate
UHDTV 8 K video	~12.5 MB/s	NHK HEVC objective

The *Multimedia over Coax Alliance* (MOCA) standard (www.mocaalliance.org) transports content over coax cables using out-of-band frequencies that don't interfere with the regular cable TV or Internet signals. The *HomeGrid* is another option for networking using twisted pair POTS (plain old telephone system) telephone connections, coax, or power lines. HomePlug offers networking through the power wiring in a home. Chapter 11 gives a list of these networking standard groups.

Some of these options have better prospects than others, but it is likely in the highly scattered consumer electronic market that more than one networking options will survive. Besides Ethernet and Wi-Fi, MOCA has support from major cable companies. This could help in pushing for their adoption. There are also companies pushing networking over power lines.

If a home network is to be used for streaming entertainment applications, it must support data rates for the various types of streaming content; see Table 9.2. Note that the DVD and Blu-ray data rates are the maximum data rate off of the media rather than the average streaming data rate. It is possible to create non-real-time media applications that involve downloading over an indeterminable amount of time on a slow network. The downloaded data is first put into a caching storage device and then played in real time from the cache storage. IPTV may use an approach like this if streaming bandwidth isn't available and typically has some buffered content to deal with network irregularities.

The storage devices themselves operate at much higher bandwidth than the networking infrastructure. A 3.5 inch hard disk drive used in a DVR, for instance, commonly provides a 50 MB/s data rate. Thus, the network connections are usually the bottleneck for entertainment content or data transfer throughout the home.

The availability of greater processing capability at the content creators as well as in consumer devices has enabled more complex compression technologies that reduce the required bandwidth for content delivery. The development of higher content compression technologies such as HEVC (requiring much more intense encoding processing) is a key enabler of delivering 4 K content to consumers. HEVC holds the promise of up to a 50% gain in compression vs. conventional H.264 compression.

In January 2013, the HEVC standard (MPEG2 H.265) was announced[1] and demonstrated at the NAB show. HEVC is a compression standard developed jointly by ISO/MPEG and ITU-T/VCEG (video coding expert group). It does this by lossy compression, where the compressed content is governed by an algorithm for minimizing the subjective impact of that compression. The estimated 2:1 compression requires significantly more processing power for encoding (this may be 100× conventional format encoding processing) as well as decoding. The additional encoding processing is well within current generation equipment and HEVC compression, and similar approaches are often used for 4 K content distribution.

In practice, a T-100 Ethernet network rated at 100 Mbps runs at about 10 MB/s including the effects of network overhead. It would be difficult to handle multiple high-definition media streams while still handling home data. For instance, if 10 Mbps data (1.25 MB/s) is required to support home data transfer requirements, the T-100 network could only handle at most 2–4 HDTV video data streams. The network could handle 4–10 SD quality video streams. Many DVR/PVR suppliers can handle several simultaneous recording and play out streams at once.

A gigabit Ethernet network would allow roughly 100 MB/s after overhead, while the Multimedia over Coax Alliance (MoCA) networking scheme allows at least 34 MBps data rates. Network data rates of at least 30 MBps can support a 10 MBps general data transfer plus 35 HDTV streams and several digital music streams. A T-100 Ethernet network could support SDTV or HDTV quality video formats on a MoCA network. However, for UHDTV content and some other home traffic, a HomePlug or HomeGrid power network or a 1 or 10 gigabit Ethernet network will be required. Note that the HomePlug network would also combine the network connection with the device power plug-in, adding some extra convenience for the consumer.

---

[1] New video codec to ease pressure on global networks, ITU Press Release, Geneva, January 25, 2013.

9.4 Push Vs. Pull Market for Home Networks

Most of the growth of home networking has been based on retail rather than a service sales model, except for apartments and other concentrated living units. Most home networks are installed and maintained by the individual homeowner or, in some cases where the homeowner can afford it, by a specialized home network specialist or by the cable or DSL company that provided the home connectivity. Originally consumers used a network to share hardware resources, such as printers. Today many consumers use home networks (most based upon Wi-Fi) for streaming entertainment content as well.

Additional growth in home networking will be driven by entertainment streaming, cloud-based services, and local network storage to serve smart connected devices (IoT). Entertainment and information are the big drivers today, and many of the connection suppliers are cable companies or DSL phone companies. Many televisions and setup top boxes for the home allow consumers a wide choice of content through online services such as Netflix or Amazon.

In the case of most cable providers, service, including installation of the leased equipment in the home, is the done by the provider company or its contractor. When the service provider supplies the hardware, it increases the available market by making it easier for less technically savvy consumers to own and operate additional services. The result is a significant incremental source of revenue to the service provider.

Even more important is that service providers find that offering increased services such as DVR/PVR capability in their set-top boxes increases their retention of customers, especially the most lucrative customers that demand and buy these additional services. With the rise of network DVR services and settlement of legal issues that slowed the growth of these services, network connection providers can provide these services with even lower costs (particularly service costs to support equipment in the home).

Current US cable households are somewhat more than 118 million today with the majority of these having digital cable. Cable modems supply the highest percentage of broadband Internet connections (about 60% of all broadband connections are through cable). With these overwhelming statistics, it is natural to project that coax cable could be used within the home for media and file sharing as well as being the predominate source of external home networking through cable modems. However, there is greater familiarity with Wi-Fi than cable for connectivity since it doesn't require physical wires in the wall of a home.

Another interesting trend is that about 25% of homes in America have eliminated their connection to traditional cable TV services. These so-called cord cutters are using over-the-air TV broadcasts (about 17%) or Internet-only TV services (6%) to supply their entertainment needs. These trends are higher for younger consumers and some demographic populations. For 18–34-year-olds, 22% use broadcast-only reception and 13% get videos on their TVs using the Internet.[2]

[2] http://variety.com/2016/biz/news/cord-cutting-accelerates-americans-cable-pay-report-1201814276/

9.5 Home Networks for Media Sharing

There are several approaches to create a working home network. The clear winner has been Ethernet and especially Wi-Fi. Wi-Fi is now capable of significant data rates that can handle heavy content loads, such as high-definition and even UHD content. Wi-Fi doesn't require running cables through your home to get connectivity and is thus used where cables aren't built in or where it is more convenient to use Wi-Fi than cable.

Internet and video content is brought into the home originally through cable or DSL phone lines. These external inputs bring content into the home where it can be distributed inside the home using coax (if cable is used) or, increasingly, with wireless communication.

An approach for home networking that would appeal to a service model from the cable providers is using the existing coax networks in the home. This has been a popular approach for cable companies serving customers.

However, wireless networks in the home are used by some services, e.g., satellite TV companies. Many homes have existing cable wiring, but it is not clear if this will be common in future homes. The convenience of high-speed wireless connectivity between a central distribution unit in the home and possibly the development of home IoT wireless gateways for further generations of connected smart home devices could make Wi-Fi the dominant local network connectivity approach in the near future.

As shown by the Multimedia over Cable Alliance (MoCA) (which includes members such as Arris, Broadcom, Comcast, Cox, Echostar, Intel, Taraband, MaxLinear, Technicolor, and Verizon), it is possible to build a network using the excess bandwidth available in existing coax cables in the home to transport digital content and data at up to at least 34 MB/s. This is more than three times faster than T-100 home wire-based Ethernet networking technologies. But it is about three times slower than a 1 Gb Ethernet network. It is also about four times faster than the slowest Wi-Fi and four times slower than the fastest existing Wi-Fi (of course actual Wi-Fi speeds depend upon distance and interference).

Many cable service providers use this out-of-band (that is not using the same bandwidth that the digital cable signal uses) networking technology for home digital entertainment streaming and file sharing within the home. This is used as a way to share set-top box DVR/PVR content to TVs in the home. DVR/PVRs are often network devices in the home and share their content through the cable network with other PVR/DVRs boxes or with home media players. Figure 9.1 shows how multimedia content can be shared through the home using MoCA technology.

MoCA uses 50 MHz channels in the 850–1500 MHz frequency range. Within the physical layer, orthogonal frequency division modulation (OFDM) subcarriers are adaptive-modulated by modulation schemes ranging from binary phase-shift keying (BPSK) to 256 quadrature amplitude modulation (256QAM). The medium access control (MAC) layer uses a distributed mesh network architecture with time division multiple access (TDMA) for scheduled

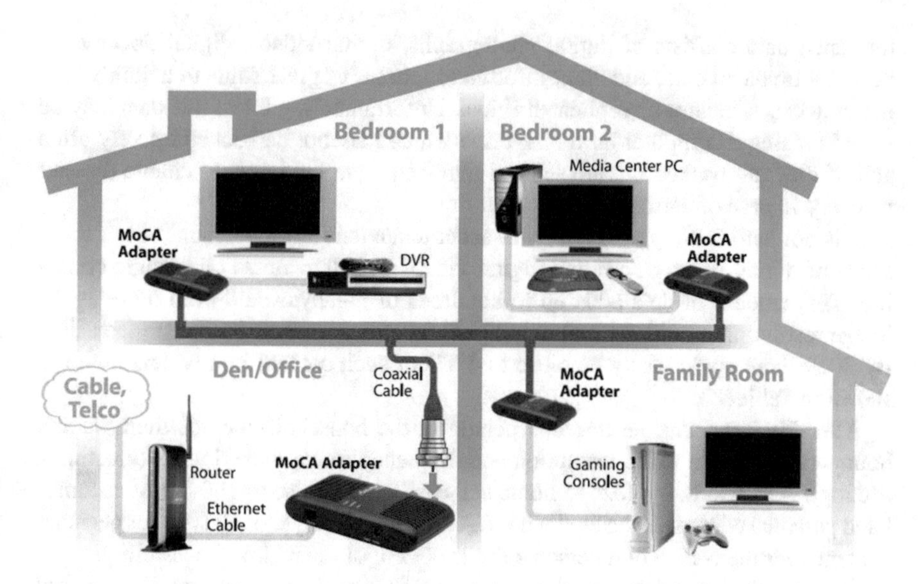

Fig. 9.1 MoCA model for home media network

access. MoCA works with all existing cable systems and is one of the most common ways that consumers connect their homes to broadband services.

MoCA was initially established in 2004. Field tests of the MoCA technology reported on April 4, 2005, at the National Cable show demonstrated that at least 100 Mbps data rate was delivered to 95% of the coax jacks tested. Multiple tests were performed in over 200 homes and multiunit dwellings validating that the MoCA technology met the requirements set in the market requirements document approved by the MoCA Board of Directors. In 2006 MoCA introduced its 1.0 specification. MoCA 2.0 was introduced in 2010.

In April 2016, the alliance introduced the MoCA 2.5 spec to support coax-based, in-home throughputs of 2.5 Gbps (~312 MBps). MoCA 2.5 is backward interoperable with MoCA 2.0 and MoCA 1.1. CableLabs documents indicate that future MoCA 3.0 technology could require supported data rates of between 2.5 Gbps and 10 Gbps (1250 MB/s) and become available in 2 to 4 years. That would put MoCA 3.0 in line with the kind of access network capacity envisioned by DOCSIS 3.1 (up to 10 Gbps downstream and at least 1 Gbps upstream).

9.6 Home Networks for Home Reference Data Backup

Once coaxial or other home networks for media sharing become popular, it is only a matter of time before these networks are also used for other networking. General data sharing would use more generalized IP over coaxial technology. Home computers are becoming the repository for gigabytes of home reference data. This home reference data consists of digital photographs, digital videos, digital documents, personal financial data, and other information that is of great value to a family and, in many cases, cannot be replicated if lost. Unfortunately, a lot of this data may be stored on single computer hard disk drive which may not be backed up very often and, if they are backed up, are seldom removed from the home to enable disaster recovery in case of earthquake, flood, or fire.

It is not hard for a typical family to accumulate tens and even hundreds of thousands of digital photographs and digital videos over the course of a single child's life. This amount of data adds up to hundreds of gigabytes and even more as the image resolution of still and moving digital records continues has been increasing. By 2019 there could easily be close to 4 TB of such content in a typical home as shown in Table 9.3.

After 2019 we assume that one person in the household records their waking hours with a certain video resolution and frequency for the next 5 years (creating a life log). As a result, by 2025, a home can easily have close to 190 TB of accumulated personal content associated with it. A personal petabyte of all types of personal content over the course of a person's life is not out of the realm of possibility.

How can these precious records be preserved? If there are stored on a single hard disk drive, the risk is too high that years of irreplaceable records can be lost. The only way to preserve this family reference data is to back it up and have multiple copies of the records available either in the home or in the cloud or both.

Home networks will be an important element in backing up content locally or into the cloud. Although it is not now a big driver for home networks, we are very close to home backup being a crucial part of the home storage environment. This backup of content can be done locally using USB or Thunderbolt interface external storage devices or with network-attached storage (NAS) devices attached to the local network. The performance of USB and Thunderbolt interfaces was examined in an earlier chapter.

Ethernet-based NAS devices are available from many external storage suppliers, including Seagate Technology, Western Digital, Toshiba, Buffalo Technology, and many other companies. Some of these devices can be set up to automatically back up multiple devices in the home. Likewise, the direct-attached external storage devices can be set up to automatically back up the device that they are connected to.

Cloud storage is available from many companies that allows consumers to back up computers and other devices to one or more cloud-based storage services. The cost of all these data backup options is declining with time, making it easier for consumers to keep more of the content that they capture and create.

9.7 The Home Internet of Things

Smart connected devices are showing up in many consumer devices and appliances. These products can provide many services to consumers including security, energy management, safety, entertainment, and various other home and consumer services.

Table 9.3 Projections for growth of a typical tech-savvy home to 2025

Annual generation of home reference data										
Year	2016	2017	2018	2019	2020	2021	2022	2023	2024	2025
Photographs										
Average storage/picture (GB)	0.003	0.003	0.003	0.003	0.003	0.004	0.004	0.004	0.004	0.004
Number of pictures per year	1569.2	1726.1	1898.7	2088.6	2297.5	2527.2	2780.0	3058.0	3363.7	3700.1
Annual photograph capacity (GB)	4.5	5.2	6.0	6.9	8.0	9.3	10.7	12.4	14.3	16.5
Home video										
GB/hour	10.8	11.3	11.9	12.5	13.1	13.8	14.4	15.2	15.9	16.7
Number of hours	34.8	38.3	42.2	46.4	51.0	56.1	61.7	67.9	74.7	82.1
Annual video capacity (GB)	375.4	433.6	500.8	578.4	668.0	771.6	891.1	1029.3	1188.8	1373.1
Documents										
Average capacity/document (GB)	0.001	0.001	0.001	0.001	0.001	0.001	0.001	0.001	0.001	0.001
Number of documents/year	62.8	69.0	75.9	83.5	91.9	101.1	111.2	122.3	134.5	148.0
Annual document capacity (GB)	0.06	0.07	0.08	0.09	0.10	0.12	0.13	0.15	0.18	0.21
Life-log/personal diary/PMA										
Sampling rate (seconds between samples)					300.0	255.0	216.8	184.2	156.6	133.1
Sampling time (seconds)					30.0	33.0	36.3	39.9	43.9	48.3
Resolution (GB/s)					0.002	0.003	0.005	0.007	0.010	0.015
Life-log accumulated capacity (GB)					3154	6122	11,883	23,068	44,778	86,922
Total capacity per year (GB)	380	439	507	585	3830	6903	12,785	24,109	45,981	88,312
Cumulative capacity (GB)	*2396*	*2835*	*3342*	*3927*	*7757*	*14,660*	*27,445*	*51,554*	*97,536*	*185,848*

These devices may have local hubs that they interact with to aggregate information and make real-time decisions. They may also interact with online cloud services. Often these devices can be accessed and controlled through a consumer's smart phone or tablet device using downloaded applications.

Major companies are offering products and services for consumer IoT applications, including Amazon, Apple, Google, Honeywell, Microsoft, and several other companies. Many of these products offer online services tied to devices in the home. These services include entertainment streaming, security, lighting and temperature control, and ordering goods or services. Standards are being developed that are making it possible to develop a distributed device ecosystem. This is a rapidly developing market, and it is revolutionizing the way consumers manage their homes and interact with merchants.

Cloud-based consumer services can use high-end servers that can implement big data analytics and machine intelligence. These capabilities can enhance the performance of local applications. In a sense many consumer applications, including current IoT devices, are a dependent upon a hybrid of local and cloud hardware and software.

Some artificial intelligence capabilities are also being built into consumer products. For instance, most smart phones have red eye removal and face and movement recognition capability built into their electronics. These devices will become even more capable of making decisions about who is in images and what to do with data as the electronics and software become even more complex. In the future, both in consumer devices and in the cloud, we will be assisted by smart services that can learn from our behavior and provide what we need, when we need it.

In order to manage the large amount of data that IoT devices for consumer applications will generate and also to provide some level of protection for consumer content, home IoT gateways could become common. These would provide local networking and real-time decisions using *artificial intelligence* (AI), while cloud services could receive summary results from these gateways, perhaps with the actual source of the data masked so it is anonymous. This could be a basis of future home privacy.

9.8 Projections for Home Network Storage

As shown in Fig. 9.2, in 2007 homes typically had at least three different communication networks which were more or less isolated from each other. These included a computer network and a cell phone or two connected to a cell phone network and an entertainment network (cable or satellite).

By 2017 these isolated networks have largely collapsed into a single network (or close interactions between networks, allowing them to work together). These unified services reduce the costs to provide these services to consumers. In addition, consumers often do not have any idea where the data used by these services actually lives. Data is collected locally and may be processed locally or in the cloud. The results of this data may be partly stored on local devices (especially where real-time responses are needed) or in the cloud.

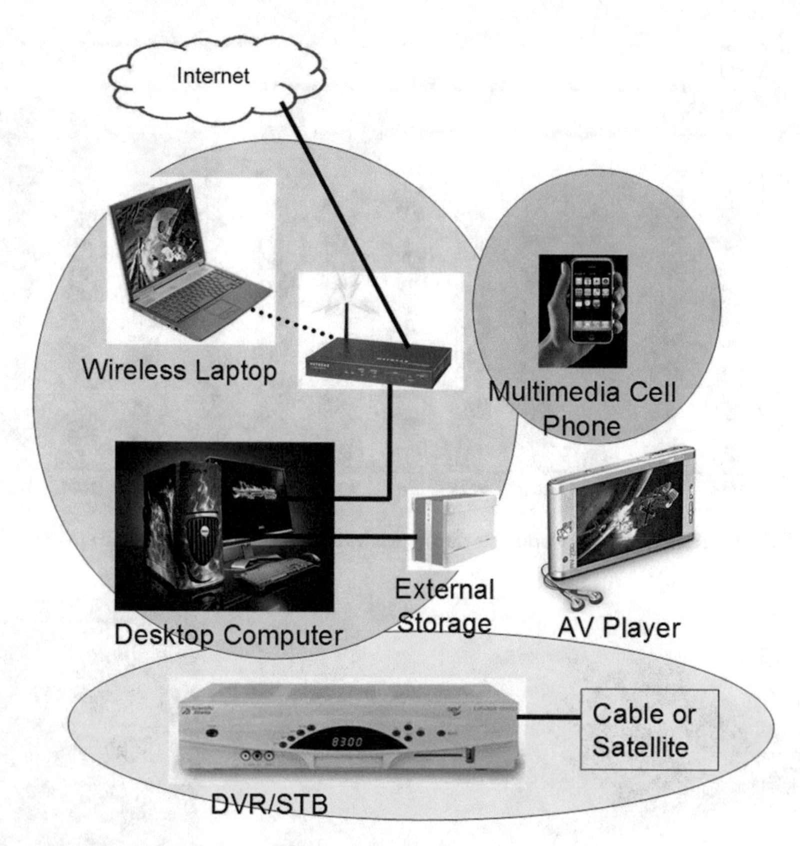

Fig. 9.2 2007 digital home with isolated home networks for computers and Internet access, communication, and entertainment

In many regards, storage for consumer data is increasing virtualized; it tends to be stored automatically where it is needed and for cloud-based copies of content, and the data can be supported and managed by common data center practices. These practices include automatic data replication, backup (both on-site and off-site), remote access and sharing of content, organization and indexing of the content, and data synchronization between devices and locations.

Storage in the home often includes external storage devices, which may be directly connected to computers, DVRs, and other products or which may have an Ethernet interface and provide shared access to digital storage assets, a network-attached storage (NAS) device. Figure 9.3 shows projections for the worldwide growth of direct- and network-attached storage devices in the home.

Cloud storage is the repository of much consumer data. For instance, many consumers keep their photos in the cloud. However, having a local copy of data provides additional protection of that data and allows faster access to mission critical data or access to data when the external network is unavailable. In addition, some consumers, like many businesses, want to control their own data on-site and may use their own NAS boxes to provide access in the home or external to the home.

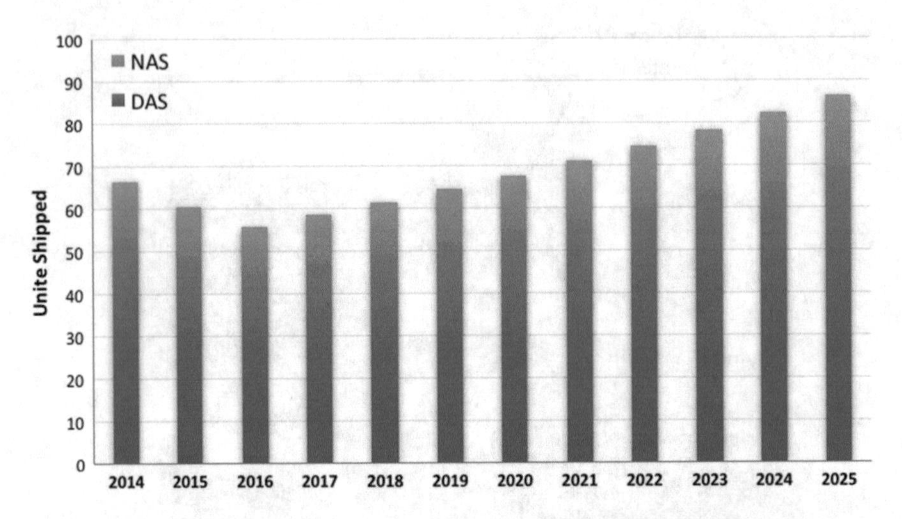

Fig. 9.3 Projected growth of worldwide direct-attached and network home storage devices

Fig. 9.4 Direct-attached and network-attached storage devices

Figure 9.4 shows some examples of home DAS and NAS storage devices. These products may contain one to five drives providing various combinations of data protection and performance (note that these can be HDDs or SSDs or a combination). Note that some of the connected devices are sold as a personal cloud storage device that you can access through the Internet.

9.9 Design of Network Storage Devices

An external hard disk drive storage device is a relatively simple device containing one or more hard disk drives, mounting hardware to keep the drive in the hardware (even if it can be easily removed), and an adapter board to bridge the Ethernet, SATA, e-SATA, Thunderbolt, or USB communications to the hard disk drive.

If there are two or more hard disk drives in the box, then various forms of RAID (redundant array of independent—originally inexpensive disks) can be used to enhance the overall system performance (such as read speed) or to increase the redundancy and reliability of the storage systems.

A RAID is a group of disk drives that appear to be a single storage device to the user with attributes that depend upon the RAID classification. For two disk drives, the two levels of RAID that can be created are called Level 0 and Level 1. Level 0 stripes the data between the two disk drives. Data striping refers to the segmentation of logically sequential data, such as a single file, so that segments can be written to multiple hard disks. Level 1 is mirroring in which the same data is written on both disks to provide good data protection from simple single drive failure mechanisms. RAID 0 provides very fast performance by the storage system but no protection of the data via redundancy. RAID 1 provides 100% redundancy by writing the same data to the two disks. As the number of disks increase, the RAID classifications often increase.

Following is a summary of RAID levels commonly used (this is close to the classification of the original RAID systems in 1987)[3]:

RAID 0

- Provides NO redundancy, since the data are written across multiple drives (so-called stripping). If one drive fails, all the data in the array will be lost.
- Provides higher data rates, since all drives are accessed in parallel.

RAID 1

- Provides data mirroring and thus high reliability. The same data is written or read on two (or more) drives.
- Faster reading, since the first drive to respond to a request will provide data, thus reducing latency.
- The cost at least doubles for a given storage capacity.
- MTBF ~ 2 M + M2/R (higher data reliability)

RAID 3

- One extra drive is added to store the parity data (error correction data). If one drive fails, the data can be recovered, and the other drives will keep working till the failed one is replaced (of course, performance will suffer).
- High reliability (cheaper than mirroring in RAID 1).
- Very high data rates. Data writing and reading occurs in parallel.
- For a given capacity, fewer drives are needed than for RAID 1.
- Controller may be more complex and expensive.

RAID 5

- Data and parity information stripping across all drives
- High reliability, high performance

[3]Patterson, D; Garth, G; Katz R. (1987). A Case for the Redundant Arrays of Inexpensive Disks (RAID). University of Berkeley, Report No UCB/SCD/87/391.

There are higher levels of RAID that mix more functions together, but a possibly more useful categorization of RAID is given below[4]:

Failure-resistant disk systems (meets criteria 1–6 minimum):

1. Protection against data loss and loss of access to data due to disk drive failure
2. Reconstruction of failed drive content to a replacement drive
3. Protection against data loss due to a "write hole"
4. Protection against data loss due to host and host I/O bus failure
5. Protection against data loss due to replaceable unit failure
6. Replaceable unit monitoring and failure indication

Failure-tolerant disk systems (meets criteria 1–15 minimum):

7. Disk automatic swap and hot swap
8. Protection against data loss due to cache failure
9. Protection against data loss due to external power failure
10. Protection against data loss due to a temperature out of operating range
11. Replaceable unit and environmental failure warning
12. Protection against loss of access to data due to device channel failure
13. Protection against loss of access to data due to controller module failure
14. Protection against loss of access to data due to cache failure
15. Protection against loss of access to data due to power supply failure

Disaster-tolerant disk systems (meets criteria 1–21 minimum):

16. Protection against loss of access to data due to host and host I/O bus failure
17. Protection against loss of access to data due to external power failure
18. Protection against loss of access to data due to component replacement
19. Protection against loss of data and loss of access to data due to multiple disk failure
20. Protection against loss of access to data due to zone failure
21. Long-distance protection against loss of data due to zone failure

A RAID can be organized on a host using a group of disks in an attached host device or with the file system located on the storage device itself as shown in Fig. 9.5. These are often referred in the enterprise storage world as software and hardware RAID, respectively. Note that because of the overhead in creating a RAID array, the total storage capacity available to the user is less than that of the sum of the individual drive storage capacities.

Although much of today's home network storage is RAID based, other approaches, such as distributed file systems, are possible that don't do RAID, although these approaches are more common with a large collection of storage devices, such as in a cloud data center.

[4] RAID Advisory Board (RAB) 1996 document.

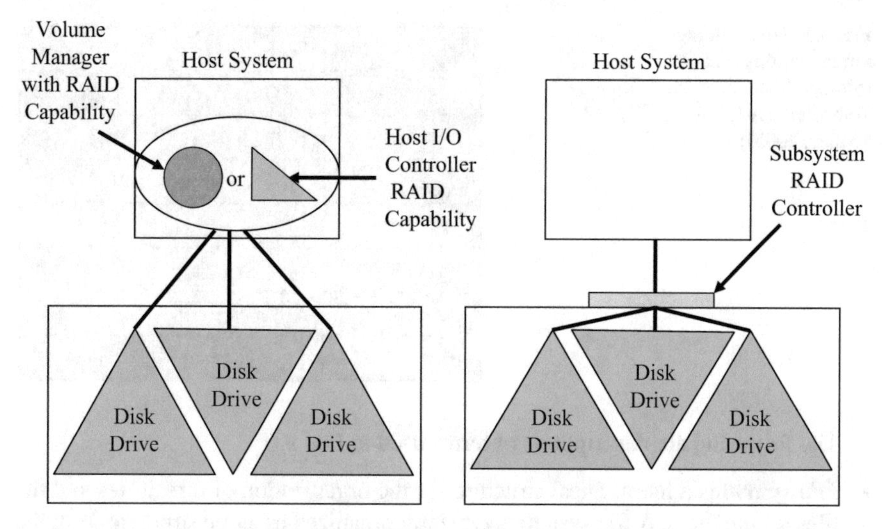

Fig. 9.5 Two approaches to creating a storage array. Software or host created RAID array (file system on the host) to the left, or subsystem created RAID array (file system on the external subsystem) to the right

9.10 Advanced Home Storage Virtualization

A RAID storage system is an example of *storage virtualization*. The physical storage devices (the disk drives) are hidden behind the *abstraction* of the RAID storage so it looks like a single storage device, such as a very large disk drive. All the other features of the array, such as redundancy to protect the data in case of a drive failure, are hidden behind this abstraction. Although an advanced user could find out how to program and modify the operation of the RAID, there is no setup required to use such a network-attached storage (NAS) device. You just plug it in, and it looks like a big hard drive to the network and to other computers and devices that access it. Such abstraction of physical devices to create virtual (or logical as they are sometimes called since they are the digital entity that can be directly accessed) devices is called virtualization.

In hardware terms, "virtualization" refers to a simplified representation of hardware resources to the user, usually to simplify hardware management (e.g., multiple devices are represented as if they are a single device to the user. Virtualization can be applied to processors, storage, and various other digital devices to hide the complexity of multiple physical devices from the average user. "Storage virtualization" refers to hiding (also known as aggregating or abstracting) the actual storage resources behind some simplified user interface.

As storage serving the home becomes more widespread and complicated (whether it is local or in the cloud), virtualization of the various storage resources for consumers will provide a way to simplify the management of the data scattered among the various storage devices and services. This becomes even more important as the number of smart connected devices in the home with some local storage increases (e.g., with IoT implementation). Figure 9.6 offers a graphic illustration of how the home storage virtualization stack may appear.

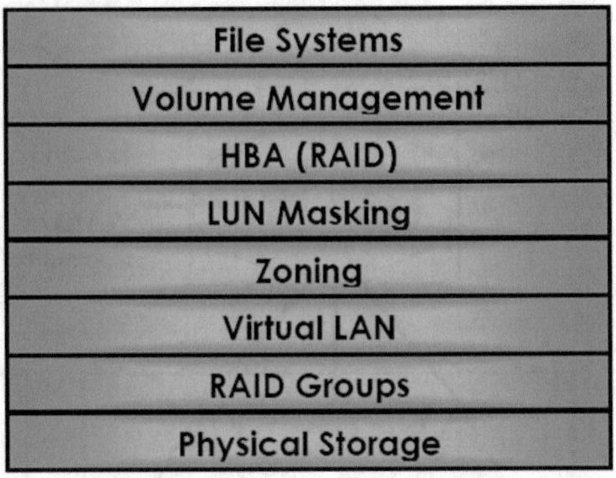

Fig. 9.6 Home storage virtualization stack (Storage Virtualization, Tom Clark (Addison Wesley, 2005))

File Systems
Volume Management
HBA (RAID)
LUN Masking
Zoning
Virtual LAN
RAID Groups
Physical Storage

The following are descriptions of terms used in Fig. 9.6:

- *File system* is a hierarchical structure for the organization of directories and the files within them. A file system is typically organized as a tree structure from the main or root directory.
- *Volume management* is a software used to aggregate or subdivide physical storage assets into virtual volumes.
- A *host bus adapter* or *HBA* is an interface between a server or workstation bus and the Fibre Channel network. Hardware-based RAIDs are often organized at the HBA level.
- *LUNs* are logical unit numbers, which are an SCSI identifier for a logical unit within a storage target. *LUN masking* is a mechanism for making only certain LUNs visible to an initiator.
- *Zoning* is a function of network fabric switches that allow segregation of storage nodes by physical port or World Wide Name.
- A *virtual LUN* is a LUN that is created that actually identifies several RAID or
- physical storage devices as a single LUN.
- *RAID* is a redundant array of independent disks that is usually used to provide data recovery via data reconstruction in the event of disk failure. A *RAID group* is a group composed of one or more RAID arrays.
- *Physical storage* is an actual storage device such as a hard disk drive.

Storage virtualization can occur at several levels. It can be virtualization or aggregation at the block or file level or at the disk level. It can be done at the host, in the network, or in the storage device or appliance. It is assumed that home storage virtualization will occur in-band (that is in the electronic transport used for the data itself). Figure 9.7 shows the types of storage virtualization that can be used for home network storage as well as the various places that this virtualization can be done.

Consumers use digital storage for various functions, and thus there can be many copies of content on computers, in the cloud, A/V media players, home servers, backup storage devices, automobile entertainment systems, and on other consumer products. Storage devices in the home or cloud storage services are becoming the repository of a family's valuable digital content such as photographs and family videos.

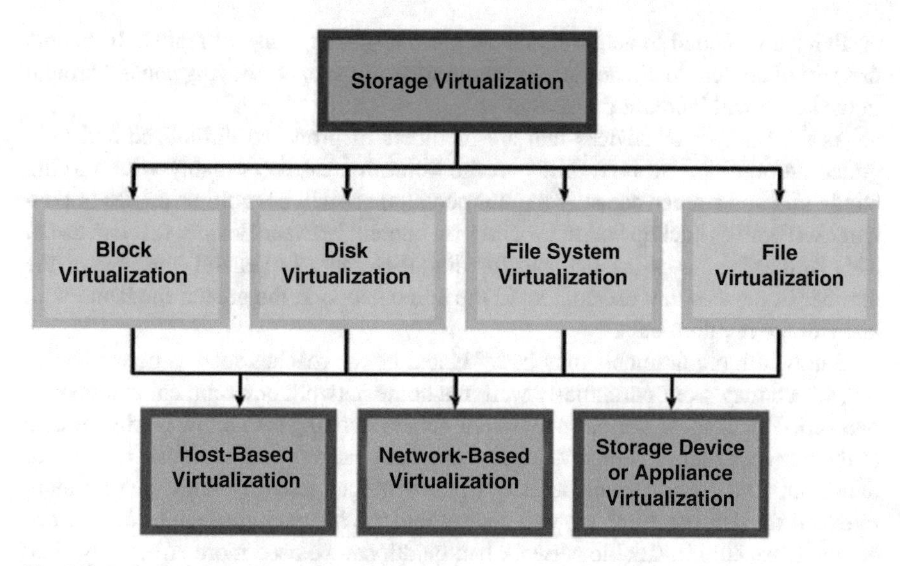

Fig. 9.7 Paths to home storage virtualization

As the storage on devices in the home grows, their number in and around the home increases. As the traffic on the home network increases, there will develop a need to extend the storage virtualization concept to all the digital storage in the entire home as well as any storage serving the home in the cloud. A whole home virtualized storage system should provide content protection, efficient use of network resources, content synchronization where it is needed, and organization of the content so it can be effectively found and used as needed and ease of use.

If all the storage and applications in a home are all managed as part of an entire ecosystem, whether provided by Amazon, Google, Apple, Microsoft, or other companies, and whether outside of a home, it may be easier to create efficient storage management, at least within that ecosystem. Between these systems, there is a need for standardization so the consumer is well served by whatever collection of smart devices and services he/she decides to use.

9.11 Home Network Storage and Content Sharing Within the Home

In Chap. 5 we discussed bandwidth requirements for a network supporting a home media center with SD, HD, or UHD content. In the future, we could expect these requirements to increase even further. Next-generation TV content with 4 K or 8 K Ultra HD will boost bandwidth and storage requirements significantly. Even current 8 K video doesn't result in fully real-life clarity and doesn't provide an immersive experience. Virtual reality and augmented reality and even higher resolution at 16 K or above are needed to achieve an even more engaging sense of reality. To handle this sort of content load, the storage requirements to support moving content around in the home will increase dramatically.

As the number of devices that are members of proposed virtualized and integrated storage or a storage utility in the home increases (probably with varying bandwidth and storage capacity requirements), there will be multiple copies of content used both as backup and to synchronize content between devices that can use it. This further increases the demand for digital storage. In general, the greater the number of devices that use content in the home network, the greater the bandwidth and storage requirements.

Bandwidth requirements may be mitigated by approaches such as network coding, which may work particularly well in a home network environment and may be part of IoT packages. The basic notion of network coding is to allow mixing of data at the network nodes. A receiver sees these data packets that contain a mixture of data from all the sources and deduces from them the messages that were originally intended for that receiver.[5] We will not get into the interesting specifics of network coding here, but if available network bandwidth can be used more effectively, then more content will be shared between devices in a network.

As more content is shared, this content must be stored on either end of the transmission either as temporary or longer-term digital storage. In a very real way, the required digital storage capacity is the physical manifestation of the content. The amount of storage can vary depending upon compression, but the basic transfer and storage of information are critical to the use of the content and thus the value of the network.

In the next chapter, we shall look at the requirements for sharing digital files in the home as well as the impact of virtual and augmented reality and social networking and other ways that people are finding new value in sharing their content with others. These services will require local as well as cloud-based services and storage, and the digital storage and bandwidth requirements will continue to grow.

9.12 Privacy, Content Protection, and Sharing in Home Network Storage

In order for content owners to be comfortable with sharing content through a home network (such as might be connected to a home media center), some form of *digital rights management* (DRM) needs to be implemented. Various schemes have been created to ensure that content does not stray from the people that have purchased a legitimate right to that content. These schemes often suffer from making use of this content inconvenient for consumers. It is clear that the proper balance needs to be established for the rights of content owners and consumers in order for the full market potential of today's technology to be realized.

---

[5] *Network Coding: An Introduction*, Tracey Ho and Desmond S. Lun, https://pdfs.semanticscholar.org/7d7f/1e4f2697ed68f49be811d326ff09607af371.pdf

Several forms of DRM are available for home entertainment networks. These include:

- The *Digital Living Network Alliance* (DLNA)—for static home devices (appears to also incorporate the older Coral Consortium)
- *Open Mobile Alliance* (OMA)—for mobile devices
- *MPEG* —also for mobile devices

On the other hand, some content owners offer music and other content with a non-DRM option making it easier to transfer and share the files between the user's devices in and around the home. These content owners usually charge somewhat less to purchase a DRM version of a music track. The *Creative Commons* offers a wide variety of licenses that can be used to allow many different use licenses for a content owner.

Personal and family content may be stored and shared from home networks or the cloud. This content must be protected from intrusion and theft as well in order to protect the privacy of the users. Various ways can be used to protect this content from unauthorized access. Ways to protect the privacy of personal content include:

- Encryption of the content to prevent access without the encryption key
- Firewalls to prevent outside unauthorized access
- Some form of digital rights management for personal content

The last approach to protect personal content is interesting in an age where more and more people are creating their own blogs and pod casts and share their personal videos and photographs (through various forms of social networking) over the Internet. As the amount of this content grows (today there is more content that relates to humans than commercial content that does not relate to humans) and as people share more and more of their content, we need some way for individuals to control access to this content and in some cases to create some way for people to get some compensation for the use of their content.

The result could be a very large micro-market for content created by individuals. These creators could allow access and use of this content by others for some small transaction fee. Approaches such as the *Creative Commons* could become the basis of such a personal content DRM, allowing some free access and use with attribution while allowing quasi-commercial access as well.

While much content will be shared between tight social groups, such as families, undoubtedly, some of this content will intentionally or non-intentionally find its way to a larger audience and some means must be in place to protect the rights of the original content creator, or if appropriate, to create some way to compensate the creator for commercial use.

With such a large supply of content available, finding shared content that fits one's needs will become increasingly challenging. It is likely that many people will turn to aggregators of such content to provide access to this material. Such aggregators could provide the mechanism for quasi-commercial exchanges and micropayments for the use of content, splitting the payments between themselves and the content creators. There are organizations that do this sort of aggregation of content today, and this trend is likely to increase in the future.

9.13 Chapter Summary

- Many options are available and being commercially developed for home networking including use of wired and wireless Ethernet as well as using existing home networks such as coaxial cable, phone, and power lines. Wireless Ethernet is the clear winner, but there are many homes that use other networking approaches as well.
- While home networking is mostly a retail market with installation by the user or a professional home installation service, this market has seen some additional offerings by service providers such as cable companies or telecommunication companies as a means of satisfying customers' total entertainment and home networking needs and so to create a sticky service offering to retain customer loyalty.
- The largest uses of home networks are for media sharing and streaming, social networking, and cloud services as well as home reference data backup.
- Today with the growth of personal data and data collected by corporations and governments about individuals, human data exceeds all other types of data. Furthermore, with the development of "life logs" to record your life as it happens, accumulated personal data will greatly exceed commercial data.
- We present a projection for the growth of direct-attached and network storage devices in the home with one or several hard disk or solid-state drives, and we discuss the design of multiple disk drive home network devices including RAID architectures.
- We discuss several technologies for DRM and privacy protection associated with commercial- and user-generated content.

Chapter 10
The Future of Home Digital Storage

10.1 Objectives in This Chapter

- Estimate digital storage capacity demand created by personal content sharing in the home and across the Internet.
- Present the advantages and disadvantages of convergence vs. single-purpose devices and the implications for digital storage.
- Study the impact of downloading and streaming on physical content distribution.
- Look at the use of a life log of ongoing recorded life experiences to create a personal memory assistant and what that enables.
- Explore the concept of distributed managed home storage pool for IoT and Machine Learning Processing.
- Find out what new sorts of devices and business opportunities may be enabled by extensive personal life databases created by life logs and the cloud.
- Show estimates for the growth of digital storage device units as well as storage capacities and the requirements for these applications in the future.
- Examine the implications of digital content as our cultural legacy and history for future generations.

10.2 Digital Storage Requirements for Home Data Sharing and Social Networking

In the home, the types of connections that can occur between devices that share content can be complex. These communications can be *point-to-point* or *multicast* (going from one source to several points). They may be wireless or wired as well. There can exist short-term and longer-term associations between two or more devices in the handling of digital content.

T.M. Coughlin, *Digital Storage in Consumer Electronics*,
https://doi.org/10.1007/978-3-319-69907-3_10

We shall create a simplified model to estimate the storage demands created by the sharing of content between multiple devices within a single home and between multiple homes connected through a network, such as the Internet and the cloud. This calculation will start with an estimate of single-device storage requirements for retention and sharing of content (including the impact of data moved to and processed in the cloud) and move on to cover the case of content sharing within the home. Finally we shall extend this model to estimate required storage capacity for content sharing between multiple homes and cloud storage.

10.2.1 Storage Capacity Requirements for Single-Use Devices

In order to create an estimate of how much digital storage capacity is required for the home and ultimately for data sharing over the Internet, let us start with individual device storage requirements. This analysis will be ad hoc, but the assumptions used for the calculations as well as the formula will be included so that the reader can modify the analysis to fit other assumptions. The described case is used as an example only. The actual numbers that should be substituted in the formulas to calculate the storage capacity for short-term cache and long-term retention for a single-use device may differ from these numbers.

We shall assume a *single-purpose device* in these calculations, and we shall assume that the content will be completely downloaded before it is used (this is a very idealized device today as most modern devices serve multiple functions and can stream content without a complete download). This analysis includes the time (t) required to download and view that content. The time t is composed of two subunits of time. One is the time for the actual download (t_d) and the other for using the content (t_u). If we multiply the data rate of content delivery (D) to the device by the download time, we get a resulting storage capacity needed for the download of content C, in Eq. 10.1:

$$C = D^* t_d \tag{10.1}$$

If content is used, once acquired, the minimum time it would remain on the device is the total time $t = t_d + t_u$. Thus the minimum storage capacity needed for this content during the total time t is C. Depending upon the richness of the content and how long until the intended use, t_u may be longer or shorter than t_d.

As an example, if we have a connection between the single-purpose device and some outside source of content that provides an available data rate to the device of 1000 Mb/s (e.g., 1 Gb Ethernet at ~100 MB/s) and the resulting content has a capacity of 375 MB, then it will take about 3.75 s to download the content. If the content is then used immediately, playing out at a data rate of 0.25 MB/s (an SD video streaming rate), then it will take 1500 s (25 min) to play out this content. Thus the total time for download and playback takes about 25 min (assuming that we don't start playing until this download is completed and we don't delay in playing the content). With

Table 10.1 Comparison of stored object size and annual accumulation for a single-use device with various types of digital content for 100 MB/s downloads for 2 hours per day

Content	Playback data rate (MB/s)	5-min content size (GB)	t_d (s)	t_u (s)	Annual data retention (%)	Annual data accumulation (GB)
MP3 audio track	0.02	0.0036	0.04	180	0.2	1.1
SDTV video	0.25	0.075	0.75	300	0.2	13.1
HDTV video	0.625	0.188	1.9	300	0.2	32.7
4K UDHTV video	1.88	0.563	5.6	300	0.2	96.8
8K UDHTV video	3.13	0.938	9.4	300	0.2	159.3

these assumptions, we require the single-use device to have a data storage capacity of at least 375 MB for 25 min. This may be temporary cache storage which will be recycled for other uses, unless the user decides to keep this content for a longer time.

If we assume that 2% of all the content downloaded is kept and that the device is used as described for 2 hours per day, every day, then the retained content for this single-use device over the course of a day is (7200 s)*(375 MB/1504 s)*0.02 = 35. 9 MB. Over the course of a year, this single-use device would be responsible for the accumulation of about 13.1 GB of content. The actual amount of downloaded content kept will vary, user to user.

We can carry out the same sort of calculation for different sorts of content as shown in Table 10.1, assuming the same usage conditions and channel data rate. In the table, we assume typical MP3 tracks of 3-min and 5-min videos of SD, HD, and 4 K UHD content.

Devices that are used for personal content retention such as digital still and video cameras will have much higher retention rates (approaching 100%), but the daily and annual usage may or may not approach the 2 hours per day calculated here. If people in a home start to create life logs with actual recording of their daily waking lives, then the storage requirements will be much higher.

Many devices perform multiple functions and so their storage requirements will be different than those calculated here—probably higher if they combine video with audio content. Also, the long-term content storage may not be in the device itself if it is part of a storage network or if the content is delivered by a cloud-based application or service. Long-term retention in this case could be within a network-attached storage device or in the cloud. Longer-term retention could also be on a direct-attached storage device that is not part of a network. In general, the calculations in this section are meant to indicate that the desire to retain even a small amount of content after it is acquired (in this case 2%) will result in significant long-term storage capacity accumulated demand.

10.2.2 A Model Home for Data Sharing

In an actual home, we can assume that there are a number of effective single-use devices, even if most of the actual devices have multiple uses (the multiple-use devices consist of several single-use devices in this model). The number of such single-use devices requiring either short-term cache storage and/or access to longer-term storage capacity will probably increase with time as such devices become more and more a part of our everyday lives. It is a central thesis of this book that consumer devices with digital storage will have to communicate with each other in order to protect personal and commercial content from loss, in order to effectively organize and use that content and to share and synchronize content. In this section, we examine the storage capacity implications of networked storage or cloud storage supporting a single-home environment.

As we estimated in the section above, the digital storage capacity requirements for a single-use device in the home is given by Eq. 10.1. In an actual home, there may be a number of different devices with different storage requirements for content. We will create a model home with a number of single-purpose devices requiring local content storage. In this home, we assume two adults, two children, and a pet. The two adults share a desktop computer with a 1 TB hard drive and a laptop computer with a 500 GB hard drive or solid-state drive and the two children each have a laptop computer with a 256 GB SSD. Each adult and each child has a smart phone with a digital still and video camera having 32 GB flash memory and which is also used for email. In addition, each adult and child has a tablet with 64 GB of flash memory capacity that is used for web surfing, watching videos, and reading e-books (each of these functions will be treated as a separate single-purpose device.

In addition, the family has two digital still cameras with 12 megapixel resolution with 64 GB flash memory cards used in each. They also have a video camera with a 256 GB internal storage capacity (could be HDD or more likely SSD). The family has a direct-attached external storage device with 5 TB used to back up the desktop computer and to contain some personal content, a 1 TB external storage device for backing up the laptop computers and a 10 GB network storage device for secondary backup of the other storage devices as well as content sharing. In addition, the family has 100 GB of cloud storage. The family also has a TV attached to a cable or satellite network with a set-top having 500 GB storage capacity.

The devices and the assumed storage capacity in this model home are summarized below:

- Desktop computer with 1 TB hard drive
- Laptop computer with 500 GB hard drive
- Two laptop computers with 250 GB SSDs
- Four smart phones with 32 GB flash memory
- Four tablets with 64 GB flash memory
- Two digital cameras with 64 GB removable flash memory
- Digital video camera with 256 GB memory (HDD or SSD)
- One external direct-attached storage device with 5 TB storage capacity

- One external direct-attached storage device with 1 TB storage capacity
- Network-attached storage device with 10 TB HDD storage capacity
- TV with set-top box having a 500 GB HDD

The total raw storage capacity on these devices in and around the home is about 19,268 GB or about 19 TB spread across 19 devices, many of these being multiple-use devices. This raw capacity does not include storage capacity on spare memory cards or commercial content on CDs, DVDs, or Blu-ray Discs. The home network assumed in this model is an 802.11a/c Wi-Fi network with about a 100 MB/s data rate. This model reflects the common state of the art for today's home storage networks.

In enterprise storage systems, 70 + % total storage utilization is common today. In the case of home storage device utilization, this varies depending upon the device as well as personal practices, supporting services and home economics. Content capture device storage is liable to be emptied regularly into longer-term storage devices, so devices such as digital still and video cameras, including those on mobile phones, are liable to run at a low average utilization rate (let's say 10%). Tablets and smart phones are added to and subtracted from on a regular basis, let's say they run at 60% storage utilization, and the content on these devices is primarily a copy of content on one or more of the personal computers and/or the network storage device, or is temporary streamed or downloaded from the cloud.

The personal computers will be assumed to run at 70% of total capacity. The direct-attached external storage devices are used for backup of the computers that they are attached to as well as containing some unique personal content. We will assume that they run at 70% capacity. Likewise the network-attached storage device might contain an additional backup copy of personal content from the computers and DAS devices as well as some historical personal content that can be shared between the computers and other devices on the network. We will also assume 70% utilization for the network storage device. Because the set-top box is constantly recording and filling its drive (while overwriting old content) we assume that this is 100% utilization.

Table 10.2 shows the assumptions in this home model including content on the network that is commercial (such as purchased software, images, music files, and video files as well as personal content such as still and video images created by family members, computer files (text documents, spreadsheets, presentations, etc.), and any other content that originates from the home or family members. The model shows raw storage capacity, net capacity taking into account utilization, first instance commercial capacity, first instance personal capacity, and copy capacity (both commercial and personal). *First instance content* is the original copy of commercial or personal content which may be additionally copied on other devices for data protection or for local access when away from the network.

In this table, commercial content such as a Blu-ray Disc or DVD, CDs, and other contents released on optical disks that cannot currently be legally downloaded into a home network are not included in this model and would likely add more than a terabyte of commercial content to that in Table 10.2. Note that this could change

Table 10.2 Summary of assumptions about model home storage (storage capacities are in GB)

Device	Device	Raw capacity	Utilization (%)	Net used capacity	First instance commercial capacity	First instance personal capacity	Copied capacity
Desktop computer	1	1000	70	700.0	340.0	360.0	
Laptop computer	2	500	70	350.0	183.3	166.7	
Laptop computer	3	256	70	179.2	64.0	115.2	
Laptop computer	4	256	70	179.2	64.0	115.2	
Smart phone	5	32	60	19.2	1.0	14.2	4.0
Smart phone	6	32	60	19.2	1.0	14.2	4.0
Smart phone	7	32	60	19.2	1.0	14.2	4.0
Smart phone	8	32	60	19.2	1.0	14.2	4.0
Tablet	9	64	60	38.4	2.0	4.4	32.0
Tablet	10	64	60	38.4	2.0	4.4	32.0
Tablet	11	64	60	38.4	2.0	4.4	32.0
Tablet	12	64	60	38.4	2.0	4.4	32.0
Digital still Camera	13	64	10	6.4		6.4	
Digital still Camera	14	64	10	6.4		6.4	
Digital video camera	15	256	10	25.6		25.6	
DAS ext. storage device	16	5000	70	3500.0		1166.7	2333.3
DAS ext. storage device	17	1000	70	700.0		280.0	420.0
NAS storage device	18	10,000	70	7000.0		2529.7	4473.60
ST box	19	500	100	500.0	500.0		
Total		*19,268*		*12,867*	*1163.3*	*4846.3*	*7370.9*

dramatically with the widespread use of *controlled access* and other DRM standards that allow and support purchase of network rights for the home. With network rights, downloading of optical disk content as well as Internet content to a home network would enable sharing of this content between various devices in and around the home. Cloud-based content now provides ways to provide controlled access content for the home and personal entertainment devices.

As stated elsewhere in this book, it is expected that total home content storage (including commercial and personal content in the home or accessible through the cloud) will be in the 200 TB range for many homes by the mid-2020s. Also, much of this content could be generated by the members of the household (especially with life logs).

As a replacement for local network storage or in addition to local network storage, many households keep some or all of their copied content (and even first instance personal content) in cloud storage. For instance, we might replace the 10 TB of NAS storage in the home with 5 TB of cloud storage, if the content is compressed.

Note that in this model about 112.8 GB of new personal content resides on mobile devices that has not yet been incorporated into the home network and backup system. Improvements in virtualization and management of storage assets such as those in these mobile content creation devices could copy this content more readily onto the storage network and potentially into the cloud as well. This makes this content available for the use of the household members and protects this content from loss. Total first instance commercial content on the network is about 1163 GB, while first instance personal content is about 4846 GB. Thus the total unique content that is available for sharing between devices on the home network is about 6009 GB (~6 TB).

10.2.3 Storage Capacity Requirements for Home Content Sharing Using Single-Purpose Devices

We shall tackle a calculation of the storage capacity requirements for a home storage network allowing optimal sharing of content. We draw upon the single-purpose device capacity equation above and the network value formulas from Metcalfe's and Reed's laws.

Bob Metcalfe, founder of 3Com Corporation and a major designer of Ethernet networking, stated that "The value of a network increases exponentially with the number of nodes." That is, where N is the number of nodes (devices or applications) connected on the network, the value of being a member of this connected network increases as N-squared, N^2. This is proportional to the total number of pairs of nodes in the network. This relationship is known as *Metcalfe's law*. The actual equation for the total number of node pairs is given by Eq. 10.2:

$$\frac{N(N-1)}{2} \tag{10.2}$$

where N is the number of participants.

Since sharing of digital content generally occurs between individual content nodes as in Metcalfe's law, we can use this equation in our calculation of digital storage capacity required for data sharing. However, sharing might also occur between different groups of nodes (besides just pairs). In this case, the total number of subgroups possible in the network follows *Reed's law*, given in Eq. 10.3:

$$2^N - N - 1 \tag{10.3}$$

We might equate the "value" of a network to the content shared across the network and the storage capacity used as the physical manifestation of this content, and thus shared storage capacity for content becomes a measure of content value.

We will use these formulas for a number of potential interacting entities in a network to create equations that describe possible digital storage capacity requirements due to content sharing within a network. Within a home, this shared content is usually long-term stored content in the home, that is, data that is currently on the home storage network or that is added to the home storage network with the intention of long-term storage.

In a home storage environment, for personal family content, it is assumed that there is a master copy of the content somewhere in the home network, and there may be short-term copies elsewhere in devices permanently or temporarily attached to the network that are created by sharing content. We will make some estimates of the number of long-term and short-term (cache) storage in this calculation of home network storage estimates.

At any point in time, one or more of the devices in the household will be connected to the home network. Of these connected devices, one or more of them might be sharing content with one or more of the other devices. For each of the home devices (preferably single-purpose devices in this model even if the device can have multiple uses), there is a probability of participation at a point in time $p_i(t)$, where "i" is the number designating the device and t is the time. For instance, in the model household example $i = 1$ would be the first desktop computer in whatever single-use mode it is in at the time while $i = 8$ would be a smart phone. Object "i" will be expected to be uploading or downloading content during some period of time T_i.

Due to the creation of both short-term and long-term copies of content during content sharing at a point in time, digital storage requirements could follow a form like Eq. 10.4 if content is only shared between a node pair or Eq. 10.5 if content can be shared between a larger group:

$$S = C\left(\frac{m^2 - m}{2}\right) \quad \text{Metcalfe's form (number of node pairs)} \tag{10.4}$$

Table 10.3 Calculation of shared storage capacity for content sharing in a network (GB)

Number of nodes	Participation factor	Metcalfe (GB)	Reed (GB)
10	0.2	30	30
100	0.2	5700	31,456,650
1000	0.2	597,000	4.82E + 61

Table 10.4 Calculation of shared storage capacity for content sharing in a network with reduced participation as node number increases (GB)

Number of nodes	Participation factor	Metcalfe (GB)	Reed (GB)
10	0.2	30	30
100	0.05	300	780
1000	0.007	630	3600

$$S = C\left(2^m - m - 1\right) \qquad \text{Reed's form}\left(\text{number of all node groups}\right) \qquad (10.5)$$

where $m = pN$, p is a *group participation factor* reflecting the activity of the subgroups (for the case of node sharing in the home, this is the $p(i)$ described above), N is the number of nodes, C is the average size of shared content, and S is the amount of storage created by sharing. S is the shared content in bytes and is a reflection of the shared content value.

Let's assume that we have a device that will connect for 5 min with one or more other nodes in a home using an 801.11a/c Wi-Fi network operating at 100 MB/s. The size of the shared content is then 30 GB. We have a 10, 100 and 1000 node networks, and we assume in each case a 0.2 participation factor of all the possible sharing combinations participating in this sharing. With these assumptions, we get the results in Table 10.3 for the Metcalfe and Reed approaches to calculating the total content sharing required storage capacity.

We see that for a constant participation factor for the network nodes, storage capacity increases enormously with the size of the network. In particular, in the Reed form, the growth for calculated storage value is staggering. One can only assume that as the network grows, the participation factor must decrease considerably. If we assume a shrinking participation factor of the possible sharing combinations, we get a result like Table 10.4.

The result of the lowering participation factor is that for ten nodes; we have two interacting node groups, while for 100 and 1000 nodes, we have five and seven interacting node groups, respectively. The resulting digital storage requirements are much less than for Table 10.3. Again, the Reed case is considerably larger than for the equivalent Metcalfe case.

Only some of the digital capacity shared during this transaction will be translated to long-term storage, so the net accumulated storage will be less than what is calculated here. It is pretty clear though that there may need to be large digital storage capacities to support content sharing, even within the home. For instance, with five node pairs in a 100-node network participating for 5 min with the Metcalfe equation, 300 GB are required, and if we use Reed's equation, this increases to 780 GB.

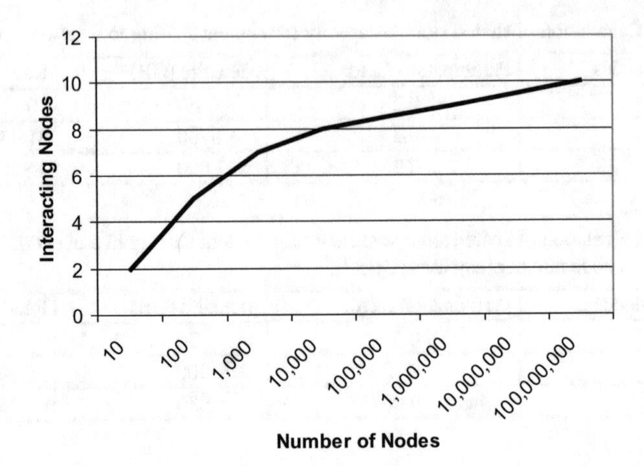

Fig. 10.1 Assumed a number of interacting nodes vs. size of network (number of nodes) for Internet sharing model

Table 10.5 Calculation of storage capacity for content sharing of 30 GB between interacting nodes per Fig. 10.1 (GB)

Number of nodes	Participation factor	Metcalfe	Reed
10	0.2	30	30
100	0.05	300	780
1000	0.007	630	3600
10,000	0.0008	840	7410
100,000	0.000085	956	10,576
1000,000	0.000009	1080	15,060
10,000,000	0.00000095	1211	21,407
100,000,000	0.0000001	1350	30,390

10.2.4 Extension of the Content Sharing Model to a Larger Network

We can extend this content sharing calculation for required storage capacity to sharing over the Internet (such as social networking or in the cloud) to find some truly large storage capacities. In order to keep the storage requirements to more reasonable level, we have assumed a network participation factor that drops significantly as the network size increases. Figure 10.1 shows the number of interacting node groups as a function of the number of network nodes (up to 100 M nodes assumed). This number of interacting node is equal to the number of nodes multiplied by the participation factor.

The resulting digital storage requirements for sharing as a function of the number of total network nodes are given in Table 10.5. We make the same assumptions for this case as for the case above with a 30 GB capacity being shared. We can see that it is very easy to come up with hundreds or even thousands of GB required to share 30 GB of capacity between parties in a sharing environment. If the participation is larger than that assumed and if it includes sharing beyond just network node pairs, then the potential storage capacity required can be very large. As in the home network sharing case, the long-term retained content will be less but over time can be quite substantial.

10.3 Integrated Multiple-Purpose Devices Vs. Dedicated Devices

Your television turns on quickly and is ready to use. When you turn your computer on, you have to select an application and go through several steps to get that application working. There has, until recently, been a struggle between whether it is better to design a device that is built to do a single function and a device that can perform several different functions. The traditional television is a hardwired single application device, while the computer can do several functions depending upon the software used.

The capability of programming a multiple-function device gives it the power to do many different things. This can be advantageous in saving the space and expense in having many single-function devices. For a mobile device, the vision of having one device (e.g. a cell phone) that serves many functions such as a phone, personal organizer, camera, etc. is very attractive.

The danger of a programmable device is that in trying to do multiple operations, it may make it difficult and almost impossible to do a single operation well. For this reason, there is a fundamental conflict between ease of use and programmable functions in a consumer device. Some devices have done a better job of easing this conflict than others. The successful devices suit the needs of many consumers and are easy for most people to figure out how to use. One example of a device that is easy for most consumers to use is the digital camera in mobile telephones. Another example, again with smart phones or tablets, is software applications that can do many different things with one device.

Cameras in mobile devices have evolved over several generations so that they are small and can take a relatively high-resolution image. It should be noted that mobile phone cameras do not yet have an optical zoom function because of the limited optics and component cost concerns for a cell phone, although this may change with new microtechnology or light wave cameras. Digital cameras have become a standard function (like those described earlier in the book) that are routinely installed in mobile phones.

The design of controls in consumer products impacts the design of products using digital storage and thus should be taken into account. Consumer products with ample digital storage are likely to have multiple functions, and thus these issues are most particularly true for these products. Today many controls are built into touch screens, making it easy to design controls matched to an application, but these controls should still be intuitive. The general objectives for user interfaces are given below:

- The user interface should be intuitive to use, that is, most people should understand how to use it just by looking at it and without additional instruction.
- The user interface should anticipate how a user will use the particular function in this device and be in the "right" location for that function.

- A user function should be enabled when (and only when) the user intends to use that function (easy to do with software applications and a touch screen).
- The user interface should be simple and easy to see or perceive in some other way, for instance, if the device can effectively use human speech, it should be apparent how to talk to it and what it understands.
- The interface should use the best interface for the application it is used for both in terms of what people are used to for that function and in terms of the fit of the interface to that function—for instance, if the device is to act like or provide control to a Blu-ray player or television, it might bring up buttons that emulate those used for these devices.
- If multiple functions for a device will conflict or make the user interface too complicated, then the number of functions should be reduced (or hidden) so that the user does not become confused (again applications allow hiding non-used functions).
- A multiple-function device that works well for users is likely to be a compromised device (although as mobile devices become more powerful, these compromises are likely to be only in terms of battery life), and a full feature set professional device is more likely to be a single-function device.

By following rules such as these, designers can make multiple-function consumer devices that use digital storage easier for consumers to use and thus easier for them to adopt into their lives. There is a big market for multiple-purpose devices that work well but don't have the full feature set as well as more professional devices that focus on providing more advanced feature sets.

Smart devices have increased in capability over the years, and today, most consumers use their smart phones rather than stand-alone cameras for normal daily photos. As a consequence, more advanced single application digital cameras with zoom functions, advanced f-stop settings, and the other things that amateur or professional photographers require sales have slumped. These products are now used primarily by professionals or semiprofessional photographers. Further advances in microphotographic technology could make smart phone cameras even more advantageous in future years.

10.4 Physical Content Distribution Vs. Downloads and Streaming

Until the mid-2000s, physical content distribution, principally using various optical media, dominated for most commercial content and personal content distribution. Mobile USB storage devices supplanted optical disks for most physical personal content distribution since that time. Many people have several of these devices that they use to carry their own content as well as copy content from other people.

With the increased availability of high-speed Internet access, downloading of music, video, and other content has become a commonplace. While this has often been done without a financial return to the content creators or distributors in the past, content owners and distributors now offer options to customers that are attractive for legitimate content purchase and distribution online. The Internet has also enabled other less formal distribution technologies such as social networking and sharing and copying music, photos, videos, etc. using web sites such as YouTube.

People are also putting their personal content on remote sites to make this material available to share with others such as family members and friends (e.g., on Facebook). This sort of content may be available using passwords or other means to control access and protect personal privacy. Earlier in this chapter, we explored the implications of sharing content outside of homes on digital storage requirements and found that social networking and other content distribution technology will create great demand for digital storage.

There may be a fundamental issue to how much content can be shared using the Internet. The Internet connection between users has only a finite bandwidth. The content that users can share in a set amount of time over the Internet can be limited by this bandwidth. This has been dealt with using intensive compression technologies on uploaded content. This is also a stimulus for intermediary online storage services that allow sharing of personal content, with the only limit being the bandwidth and time available to the person accessing the content rather than having content limitations on both sides of the exchange.

With very rich content such as raw uncompressed video, it may be that physical content distribution still makes the most sense. There is an old saw that a Fedex truck full of physical media such as optical disks or magnetic tape will have more effective bandwidth than most Internet connections and is likely to be less expensive as well.

Online distribution now exceeds physical content distribution, but this has not eliminated physical content distribution. This also fits into the current distribution models, including online purchase of such content as well as physical media such as Blu-ray Discs for very high-resolution content. Physical content, if content protection allows local working copies, such as on a home media server, can provide an effective backup of that content. Also, the content can be restored to the media server from the original copies should the media server content become corrupted.

10.5 Personal Memory Assistants

In earlier chapters, we described a *life log* that could record images, sounds, time, and location of experiences of a person during the course of his/her life. Such a life log could be the basis of a device you can carry with you that can act as a *personal memory assistant* (PMA). This device, combined with a home storage utility (described below) which can organize, summarize, and synchronize the content from a life log and a sophisticated image and voice recognition (and reconstruction) technology could create a device that can help you remember places and people that you have interacted with in the past. Such a device would act as a prosthetic memory device and could be used to help those with failing memories as well as enhancing already good memories.

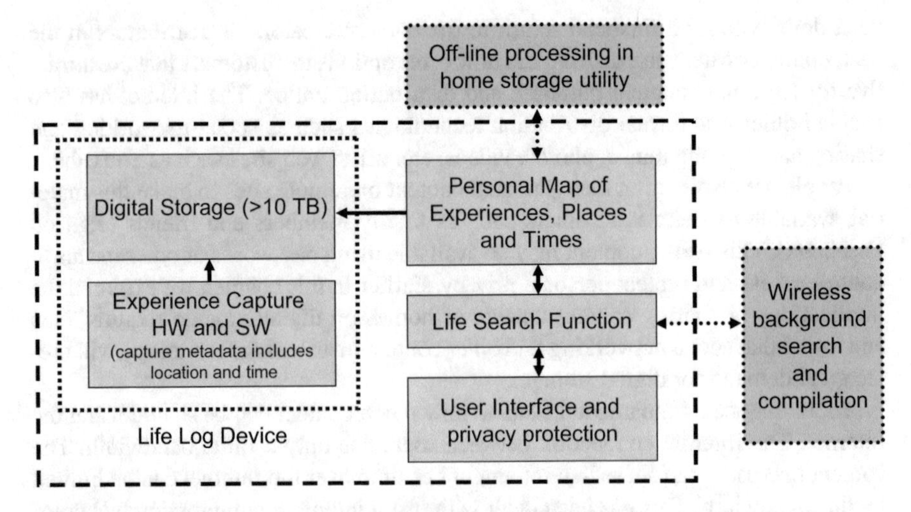

Fig. 10.2 Block diagram of personal memory assistant showing major component functions

This is a list of possible attributes for such a device:

- Storage of compressed images and sounds (and perhaps other senses as well) sampling interactions with people and locations
- Creation of a *personal map* of where you have been and when, combined with associations with these images and sounds
- The ability to search this representative database of images and sounds rapidly to find prior occurrences
- The ability to appropriately model the effects of age and time on people and places to allow more accurate identification of prior occurrences
- The ability to tie into online resources (wirelessly) to enhance the information on the person or place as required
- Rapid creation of an accurate and succinct summary of past associations and a means to share this content with the user (such as a voice that whispers in your ear telling you who the person is that you are seeing and when and where you saw them before)
- An easy to use and unobtrusive interface as well as a size, shape, and weight that make it easy to use anywhere

A PMA would require a considerable digital storage capacity as well as sophisticated computer processing and software. The storage capacity required will probably exceed 10 TB to carry sample content and metadata representing a significant portion of your life and such storage capacity (as well as the other capabilities) in a mobile device probably won't be available until about 2025 (in the original edition we thought this would be by 2015) or so. Figure 10.2 is a block diagram of such a device.

Ultimately such a personal organized database of your life and experiences could be used as the basis of some interesting entertainment applications as well as give someone else the ability to experience some of the things that made you who you are. There are obviously many privacy and other issues to be worked out to make such devices universally used, but it is my belief that the technology to make such a device is within our grasp.

10.6 Digital Storage in Everything

A key point of this book is that digital storage is a fundamental technology for the creation of modern consumer electronic devices. Digital storage is the physical manifestation of the commercial or personal content that the consumer device creates, displays, or interacts with. In this section, we will explore the implications of the consumer electronic digital storage hierarchy on the future of consumer electronics as well as how we will manage the pockets of digital storage in all the consumer devices in the home and on our person. We will also examine the threat posed by format obsolescence on the longevity of consumer content, particularly personal content.

With increased electronic integration, it is possible to integrate more and more standardized consumer applications and application functions on a digital storage device. This would save on the cost of the application, increase the application performance, and probably increase total device reliability for most applications as discussed in an earlier chapter. With a device such as a hard disk drive, this would be easiest to implement on the drive circuit board, but as electronics shrink further, these applications could also be incorporated into the circuits of solid-state storage devices such as flash memory. This intimate tie-in of digital storage and the consumer applications should also incorporate standards making communication and management of the content between various consumer devices in the home much easier.

As digital storage of various types of content is required in consumer devices, making sure that this content is managed well becomes a greater problem. The home storage utility discussed below or its equivalent, based upon cloud-based storage, could take care of this coordination. Until an overall home storage management application becomes available, people will use an ad hoc collection of applications that synchronize and back up at least some of the content. This is particularly important for content capture devices such as cameras, since that content will not be available anywhere else if the original source is lost or broken, unless it is actively managed and copies are available where they are needed.

Likewise, there is a danger to personal content associated with the obsolescence of consumer devices. Will the content on these devices be moved to a replacement device? And if there is original personal content on these devices, is it moved to a location where it can be backed up, protected, and organized? This is a growing issue as consumers have an increasing number of older devices in their garages and closets. Smart phones are the major contributor to this content for most people, and the companies providing the operating system for these devices have created cloud-based storage that helps with this migration and protection of content, but there is still much room for improvement to make sure that content on all our old devices is not lost.

For all of our consumer devices, we need efficient ways to move applications and content to new devices. If all devices could be part of a network, in many cases, a wireless network with appropriate management software for moving and protecting content as well as converting to a new device, this management is much easier.

As of today, there are no clear standards to make these functions easy, and as a consequence, users must often use semi-manual ad hoc approaches to moving content to new devices. Format obsolescence is a clear threat to the preservation of digital content and the applications that use this content. In fact, popular content formats themselves are subject to format obsolescence as more efficient compression and other technologies are developed. Over time the older formats tend to lose support, making it likely that older format content could be lost with time. If there is a large volume of older format content in many places, it can be a daunting task to move this content to the new format.

The *Storage Networking Industry Association* (SNIA) created an effort to deal with the issues around creating long-term archives. The SNIA's long-term retention (LTR) initiative is a global, multiagency group working to define best practices and storage standards for long-term digital information retention. Although these efforts are focused on long-term enterprise data retention, hopefully some of these efforts could also benefit consumers with stored content. The approach taken by the LTR is to bundle the content, its associated metadata, and the applications used to access the content in a package that is kept as one preservation element. This should make access of older content easier as the content formats and storage technologies change.

In a sense, the only way to really archive digital content for the long term is to actively manage this content, moving it to new formats as soon as they become established and keeping more than one copy in case one copy is destroyed. This is one of the functions that an integrated home storage network and ultimately a home storage pool should provide.

10.7 Home Storage Utility: When All Storage Devices Are Coordinated

In the home, there are many devices that incorporate digital storage in their design. If these devices are connected to a network, then they can be addressed by devices on that network. In a home environment, virtualization of storage could reach an even more atomic level to address issues with protecting, sharing, and organizing home and personal digital content. These networks should back up and protect user content wherever it is. We may eventually see the development of what I call a home storage pool, in which all devices connected to the network either continuously or intermittently (such as mobile devices and automobiles) would be part of a single managed storage system referred to as a *home storage utility*.

The home storage utility should provide the following basic functions:

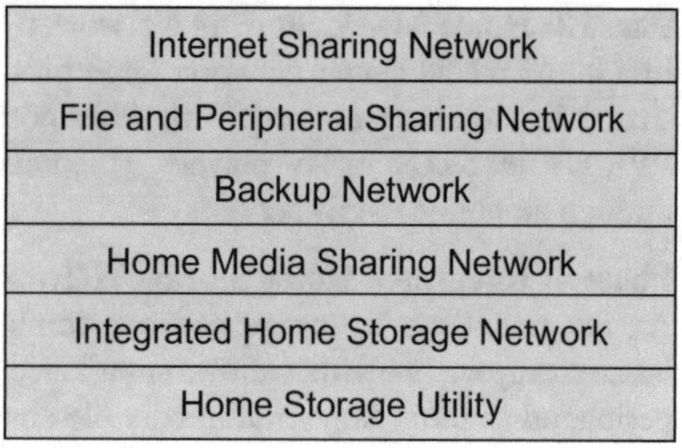

Fig. 10.3 Phases of home storage network evolution

Internet Sharing Network
File and Peripheral Sharing Network
Backup Network
Home Media Sharing Network
Integrated Home Storage Network
Home Storage Utility

- Content backup in the home including de-duplication to make better use of network and storage resources
- Content backup outside the home (to provide home disaster recovery)
- Content sharing in and around the home with optimal use of network resources
- Indexing and organizing home content so that it can be found and used when needed (particularly there needs to be an automated metadata generation facility either within the home or using cloud-based resources)
- Synchronization of content as needed (e.g., delivering music and video to A/V players)
- General management and control of storage and network resources (e.g., it should automatically identify and troubleshoot problems, either fixing or letting the owner know how to solve problems in a way they can understand)

This is a lot to ask of a home storage network, and we are not yet at this level of automated management of home and personal storage resources today. The development of home storage utility will probably occur in stages.

We have identified six phases of home network evolution, each involving ever-increasing levels of storage networking, larger content files, greater need for storage services, and eventually complete virtualization of all home storage resources. Figure 10.3 displays these phases in a graphic form.

Phase 0: Internet Sharing Networks
In the Internet sharing network, a home network is created for the sole purpose of sharing an Internet connection between two computers. This has been the major driver of home networking up to the present time. Since there is no true storage networking, we are referring to this as Phase 0.

Phase 1: File and Peripheral Sharing Networks
In a file sharing network, the connected computers and other devices on a home network share document and perhaps rich media files as well as peripherals such as scanners and printers.

Phase 2: Backup Networks
Once a network is in place with storage attached, storage devices on the network can use the network to back up their files on other storage devices on the network. Often this would occur on a specialized network backup appliance.

Phase 3: Home Media Sharing Networks
In a home media center network, large files such as those generated by video are shared in the home network with other computers or connected devices such as PVR/DVRs, STBs, or AV players. This will likely also include content managed through an online media service.

Phase 4: Integrated Home Storage Networks
An integrated home storage network combines the above features together in a home storage network that allows simultaneous Internet connection sharing, file and peripheral sharing, large multimedia file sharing, and data backup. The integrated home storage network will also include connections to fixed and mobile storage devices in the home such as those incorporated into PVR/DVR/STBs, mobile MP3, and AV players, cell phones with multimedia capability, automobile navigation and entertainment systems, and other mobile and fixed consumer devices using digital storage. In addition, the integrated home storage network will be connected to online consumer services.

As networking technology develops, it may eventually incorporate ways to more effectively utilize storage and network resources such as de-duplication to avoid making multiple copies of the same data file or a coded network[1] that allows combining multiple content into a packet that can be deconvolved by the target devices using data provided separately from each of the target devices to all the others. It may also be tied into a home security and sensor management systems and so become part of a home automation solution.

Phase 5: Home Storage Utilities
This term refers to the virtualization of all mobile and fixed device storage in and around the home. This phase of home network storage requires standard-based virtualization of the storage devices in all of the mobile and fixed consumer devices in the home. Virtualization of all the storage devices should allow easier DRM support, protection of user privacy, overall storage and content management, and reduced user and system support.

The user interacts with the utility to acquire and use any valid content. Today there exist isolated elements of these services offered by multiple service providers, but these devices are not generally interoperable and so do not include all the capabilities needed by consumers as an integrated utility package. Note that these phases outlined above can coexist and that some of them may not follow in succession.

The growth and progression of these phases are driven by three major factors. First, home networks allow sharing of files and other resources and backup of home reference data similar to practices in many businesses. Second, home networks allow sharing and distribution of entertainment content in the home. The third factor and the major motivator for the development of the last two phases is an increasing need to coordinate the various types of data and locations where that data may exist throughout and around the home.

[1] Rudolf Ahlswede, Ning Cai, Shuo-Yen Robert Li and Raymond W. Yeung; "Distributed Source Coding for Satellite Communications"; IEEE Transactions on Information Theory45(4), 1111–1120 (1999).

As multimedia and traditional files grow, it becomes harder to find and organize this content. There is a great opportunity here to develop networked storage systems that can automate the organization of home and personal content, make sure that it is where it is needed when it is needed, that it has good metadata, and that it is protected from loss.

10.8 Digital Storage in Future Consumer Electronics

As discussed earlier, large and dispersed digital storage capacity in consumer devices and in the cloud enables many new capabilities. This will also drive engineers and businesses to develop technologies that can organize and effectively use such large and fragmented content. It is important to realize that over the last 10 years and into the indefinite future, consumers have become enormous creators of content. The personal content created by these users or to support the lifestyle of these consumers will greatly exceed the amount of commercial content that will be stored in their homes and anywhere else for that matter.

As personal content becomes the majority of digital storage, this content becomes a big factor in the overall management of digital storage. If this personal content is organized and easy to access, such as that in a PMA, the result will be a material that could be used in new entertainment, educational, and communication products. A well-organized personal memory database will provide material for new capabilities such as:

- Entertainment such as games or movies could incorporate personal experiences (such as people you have met or places that you have been) to make a very personalized and more gripping experience.
- Multiple people using such a game or movie could share some of their own personal experiences to make a group experience that reflects in some way their shared experiences—whole new types of entertainment could be built upon these principles.
- Users could share life "mashups" of their daily or longer-term experiences that they could share with others.
- Business opportunities will be created based upon sharing, combining, and creating personal stored experiences, and this will be a greater business opportunity than with pure commercial content.
- A home automation system tied into elements of these personal databases could better accommodate the people that live in or visit a home based upon previous stored experiences.
- Automated systems could assist users in meeting their personal objectives and obligations based upon their personal database.

In addition to the content creation capabilities afforded by devices such as smart phones and enabled by larger and less expensive digital storage technology, the home storage pool or storage utility will play a very important role. The creation of a unified storage utility in the home should simplify content sharing and make the process or finding and using content more efficient. As people share their created content on the Internet, the role of aggregating and organizing content that is shared will become ever more valuable.

Such as huge amount of digital content will create new businesses geared to organize and protect personal content, creating new and personal entertainment and allowing efficient sharing and even monetary exchanges for the use of shared content. As personal content or content connected with people grows larger than commercial content, the economy of personal content will become the core business driver for much of the entertainment and business market.

To accomplish these new businesses, we need sophisticated automatic metadata generation based upon digital content and also drawing upon the mass of accumulated personal content and experiences that are being created in our content archives. Condensing and using this personal database will take indexing and metadata creation into new directions, ultimately making this content easier to find and more useful for the individual as well as enabling more efficient and useful sharing of this content with others.

The same increase in organization and personal information also creates new business models allowing individuals to use their personal databases in new ways to entertain and educate themselves and engage in new types of business. Ultimately, we may all use these personal databases or life logs to create virtual personal assistants that remind us who all the people in a meeting are that we have met before as well as their backgrounds and relationships as we have experienced them and by connected on online data (say using voice recognition and LinkedIn), who are the people that we don't know yet.

New standards for capturing and processing this personal data are necessary and this could be provided in the storage utility. Very large storage requirements will be needed. It is likely that by 2025, we could see homes that are associated with a petabyte of digital storage (10^{15} bytes) and most of this digital storage being used to keep personal content, copies of various personal and commercial content, and local backup of this content. Homes may also have access to terabytes (10^{12} bytes) of data remotely accessed over the Internet for remote backup or continuous data protection, which provides disaster recovery, content sharing, remote access, and synchronization between devices. Cloud-based applications will also provide entertainment, various applications and analysis, and uses of our personal content.

These services may be provided by devices sold at retail that conform to new standards enabling these functions or they may be part of a service package by a service provider such as a cable company or telecommunication company or increasingly through the cloud. In the first case, there is likely a one-time charge to purchase the equipment, and for remote access, a monthly fee, in the latter cases, for a somewhat higher monthly fee, all the hardware and software costs are included.

Service providers (including those in the cloud) have found that adding DVR and other special functionality increase the "stickiness" of their services (e.g., decreases the likelihood that a customer may switch to some other service). Service providers will find that if they provide a system with entertainment and other services well as the overall storage utility discussed here that they create the ultimate sticky business since they would provide whole house information services for their customers.

10.9 Projections for Storage Demands in New Applications

In this section, we will look at some analyst projections of digital storage needs and projected growth of different digital storage devices for consumer applications. In addition, we will make some estimates of total consumer content growth over the next few years based upon now existing consumer devices as well as projected devices such as life logs and personal memory assistants. We will also factor in the growth of storage required to support content sharing developed earlier in this chapter.

First it is important to understand the impact the age of a user has on the types of consumer devices that he/she uses. A survey by Pew Research in 2015[2] showed that younger generations tend to make the greatest use of many digital consumer electronic devices. Table 10.6 summarizes their findings.

For all these devices except desktop computers, tablet computers, and e-book readers, the younger the adult, the more likely they had the device. Also, the percentage of people who had none of these devices increased with age. We expect that for future generations, these trends will continue with each generation of younger adults being more likely to use popular consumer electronic products. These trends will increase overall demand for these products and for the digital storage that enables them.

Since newer consumer devices are likely to enable richer content, such as virtual reality, that consume more storage space, the size of storage in future consumer devices will increase, particularly as the cost of that storage to the OEM declines.

Figure 10.4 gives the history and projections for unit growth of hard disk drives for various consumer applications out to 2021.[3] Hard disk drives no longer play a significant role in mobile devices. The majority of HDDs are used in DVRs and set-top boxes as well as game consoles. This chart doesn't include external storage devices used in the home (direct attached and NAS), which would add another 15–20 M units in 2016 and 2017.

In 2016 about 20% of all hard disk drive storage shipped (in bytes) were used in consumer applications. Total disk drive storage shipments in 2016 were about 545 exabytes. 20% of this is 109 exabytes. In 2016 total shipments of flash memory was over 100 exabytes with over 50 exabytes used for consumer applications (such as smart phones). The total HDD and flash memory shipped capacity in 2016 was about 159 exabytes (not including the cloud).

We expect that by 2020, total flash memory shipments will be about 360 exabytes, with over 100 exabytes used for consumer applications. In 2020 HDDs will ship over 1300 exabytes and if 20% of this is for consumer applications, that would be 260 exabytes. In 2020 the total shipping HDD and flash memory storage capacity for consumer applications could exceed 360 exabytes (not including the cloud), more than the total storage capacity of HDDs shipped in 2011.

---

[2] Pew Research Center's Internet & American Life Project, http://www.pewinternet.org/2015/10/29/technology-device-ownership-2015/.

[3] Coughlin Associates, 2017.

Table 10.6 % of American adults in each generation who own a device

Age	18–29 years	30–49 years	50–64 years	65+ years
Smart phone	86	83	58	30
All cell phones	98	96	90	78
Tablet computer	50	57	37	32
eReader	18	19	19	19
Desktop/laptop computer	78	81	70	55
Game console	56	55	30	8
Portable gaming	21	17	12	3
MP3 player	51	51	37	13

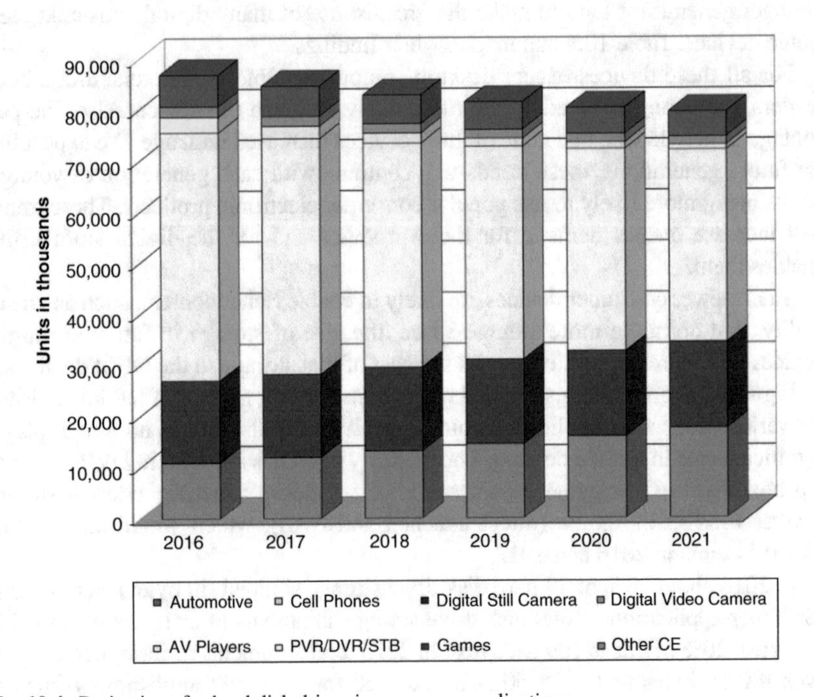

Fig. 10.4 Projections for hard disk drives in consumer applications

Optical storage media shipments are slowing in unit growth as downloading digital content has become ubiquitous. Only the high-capacity optical disks will continue to grow for the distribution of very rich 4K and eventually 8K content that would take too long to send electronically. Also, people who prefer to have a hard copy of their content (generally older people) will also be more inclined to use optical storage.

Based upon market trends in past years, we can make a projection of total exabytes shipped for consumer electronic applications. This projection, breaking down shipments of flash, hard disk drive, and optical shipment for consumer applications out to 2021 is shown in Fig. 10.5[4]. HDD and flash memory constitute the bulk of this shipping storage capacity as consumer use of optical disks declines and more content is accessed using the Internet. This is a change from the 2000s.

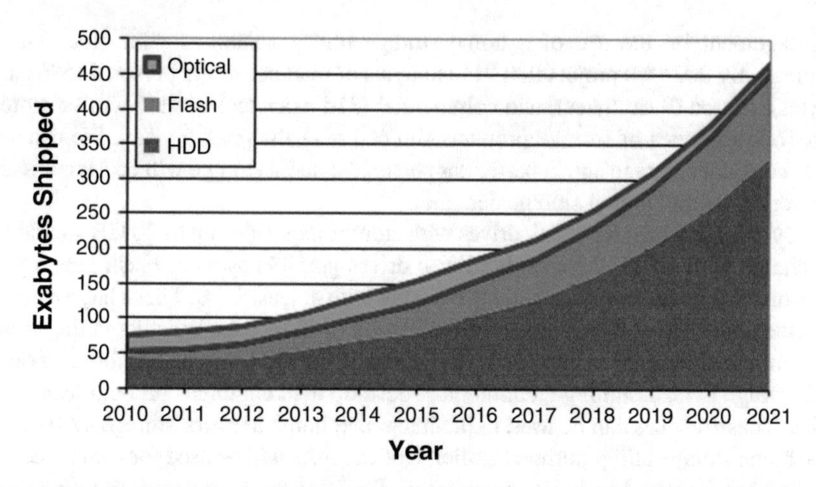

Fig. 10.5 Projections of exabytes shipped for consumer electronic applications out to 2021 including optical disks, hard disk drives, and flash memory

Table 10.7 Estimated growth of personal and commercial content stored or associated with consumer static and mobile consumer devices (storage units in exabytes)

Year	Commercial content	Self-generated personal content	Shared personal content	Total
2011	39	19	0	58
2012	59	25	0	84
2013	82	34	0	116
2014	108	44	0	152
2015	136	56	1	193
2016	164	72	1	237
2017	195	91	1	287
2018	224	114	2	340
2019	256	141	4	400
2020	285	295	6	585
2021	316	586	9	911

The effects of social networking and the growing collection and sharing of personal content as well as growth in the use of life logs and other such memory-intensive products drive storage demand (especially for hard disks and flash memory) at a much higher rate than in the past.

Next, we make an estimate of the total storage capacity required for personal and commercial content in consumer environments (both static and mobile CE products). We shall also include a factor for digital content due to shared noncommercial content. The results of this estimate are given in Table 10.7.

By 2021 we project that the total consumer storage utilization will rise to 56% (from 16% in 2011). This improvement in consumer storage utilization is primarily driven by the estimated growth in life log content in the last few years of this projection as well as increased use of home storage virtualization and more efficient management by the use of a home storage utility combined with cloud-based storage. By 2021 we project that 911 exabytes of total consumer personal (586 exabytes), shared (9 exabytes), and commercial (316 exabytes) content will be stored on 1624 exabytes of storage products shipped over the prior 5 years. We can see that consumer storage and in particular personal content storage will be a significant driver for overall digital storage demand.

By 2021 3.5 inch hard disk drives with storage capacities up to 20 GB should be available with 10 TB 2.5 inch hard disk drives possible as well. Flash memory at reasonable consumer prices will be available with at least 1 TB. These large storage devices can be used to capture and store life log or PMA data. Whether in the home, in a mobile device, or in the cloud, HDDs and SSDs will form the backbone of digital storage tiers, assuming technological development continues on both technologies. These devices can be used in the integrated home network storage systems or the home storage utility outlined earlier. Optical disks will be used for content delivery where Internet bandwidth is an issue. By 2021 we might even see new non-volatile solid-state storage technologies are used in some consumer products.

Whatever else happens in the future, as long as there are people, there will be an ever-increasing demand to entertain ourselves and to store and utilize our life experiences. The human need to know more and share more will be an enormous driver for storage capacity as well as storage performance growth.

10.10 Digital Storage as Our Cultural Heritage

There is now more information available in digital form that ever existed in earlier analog manifestations. Digital storage has enabled inexpensive capture, editing, and retention of human content in new ways. The human experience of more people could survive to inform future generations that has ever been the case in human history. In a very real sense, digital storage preserves our current civilization and the human experience for future generations. Digital storage is the means by which we will pass on our cultural heritage to future generations, similar to but richer than the older forms such as books or pictures.

However, this requires that this content actually gets passed down. There are real challenges in managing and backing up content, particularly personal content in the home and making sure that this content survives to later points in time in a format and a medium that can be read. Many people don't throw older unused content away or eliminate multiple copies of content. This makes the importance of automated systems that can help people organize and protect their personal and other valuable content even more important. Without the development of such helpful robots in systems such as the home storage utility, most people will drown in a sea of their own recorded experiences.

Another important element in preserving this content for the long term is to have online copies that are available to recover and restore your content in case your home or your storage devices are damaged or destroyed. In order to make this option popular, encryption of the content to protect the privacy of your personal content will be required, which can still be used even if all of your electronic equipment is lost. This may require some biometric component in this security to allow the reestablishment of content access. The development of a personal memory assistant will also help protect the most valuable personal data by keeping a copy of at least proxy data with the user.

Taking this concept a bit further, could a collection of personal digital histories from a single point in time be used together to create some sort of representation of collective personal experiences at that time? Could such content make history more interesting to learn and lead to greater insights into large and small events? Can this be done in such a way as to protect more personal content or to make the individuals anonymous? Once collected and protected, digital histories will be part of the tools available to help future generations understand past times. If we can learn from history, perhaps the additional introspection and understanding that access to such content could make possible will lead to a greater sensitivity to the opinions and thoughts of others. Maybe this could help us be better people. It certainly will make us more responsible for our acts.

10.11 Chapter Summary

- Sharing content within the home and through the Internet will create very large storage demand. Combined with the transition of individuals from just being consumers of content to be active generators of content, the required digital storage will be enormous.
- Inexpensive digital storage as well as faster more complex microprocessors and electronics will enable the creation of life logs and consequently personal memory assistants that combine inexpensive and pervasive personal experience storage with indexing and organization in the home storage utility. This can help you recover past personal history and associated information when it is needed. Such a device will be useful for medical as well as business and personal purposes.
- In order to simplify the management and the use of digital content in the home, various paths to home storage virtualization will be developed. Over several phases, this will eventually lead to a unified home storage utility.
- The home storage utility will provide efficient management of physical storage resources in and around the home, including in automobiles and mobile devices. The storage utility will organize digital content; make sure that it is backed up in a storage in an efficient and secure manner, and make sure that the content is synchronized with the various devices, so that content is where it is needed. In the case of life logs and other intensive content, the storage utility should create a "life script" and generate automatic metadata allowing efficient finding and use of the personal content.
- The home storage utility and home networking, in general, including home media centers, should provide proper protection of commercial and quasi-commercial content as well as allow sharing while still preserving privacy of personal content.

- Younger generations are more likely to use new technology than older generations, and this will result in increased growth in digital storage as time goes on.
- Large organized databases of personal content and experiences will lead to new business models for entertainment and education. They will also lead to new challenges for long-term data retention.
- These large personal databases and content caches will also have personal and economic value as well as collectively giving future researchers greater insight into the way we lived and the things we did.

Chapter 11
Standards for Consumer Electronic Storage and Appendices

11.1 Digital Storage Standards

There are many standard bodies involved in digital storage-related standards. Design of digital storage into consumer devices benefits from being familiar with these standards or even joining these standards' efforts. Following are some resources for various useful standards.

11.1.1 ANSI T13 Committee

The Technical Committee T13 is responsible for all interface standards relating to the popular AT Attachment (ATA) storage interface. This interface is utilized as the disk drive interface on most personal and mobile computers today as well as in most consumer electronic devices.

The charter of Technical Committee T13 is to provide a public forum for the development and enhancement of storage interface standards for high-volume personal computers. The work of T13 is open to all materially impacted individuals and organizations.

URL: www.t13.org

11.1.2 CE-ATA Standard

The *CE-SATA Working Group* develops and deploys small form factor storage specifications for consumer electronic applications. The promoter companies for the initiative were Hitachi, Intel, Marvell, Nokia, Seagate, and Toshiba. The objective of the initiative was to define and support a standard interface for small form factor disk drives that

T.M. Coughlin, *Digital Storage in Consumer Electronics*,
https://doi.org/10.1007/978-3-319-69907-3_11

addresses the requirements of the handheld and CE market segments, including low pin count, low voltage, power efficiency, cost-effectiveness, and integration efficiency. The creation of the CE-ATA technology addresses these specific requirements. Note that there are no current products using the CE-ATA specification.

URL: www.ce-ata.org

11.1.3 Serial ATA (SATA) and eSATA Standards

The *Serial ATA International Organization (SATA-IO)* is an independent, nonprofit organization developed by and for leading industry companies. Officially formed in July 2004 by incorporating the previous Serial ATA Working Group, the SATA-IO provides the industry with guidance and support for implementing the SATA specification. The standardized SATA specification replaced the 25-year-old parallel AT technology with a high-speed serial bus supporting data growth over the expected future. Note that the CompactFlash CFast card specification is based upon the Serial ATA Gen II interface.

Initially SATA was designed as an internal or inside-the-box interface technology, bringing improved performance and new features to internal PC or consumer storage. Creative designers realized that the innovative interface could reliably be expanded outside the PC, bringing the same performance and features to external storage needs instead of relying on USB or 1394 interfaces.

Called external SATA or eSATA, customers could utilize shielded cable lengths up to 2 meters outside the PC to take advantage of the benefits the SATA interface brings to storage. eSATA has specifically defined cables, connectors, and signal requirements. eSATA is hot pluggable. eSATA is not very common anymore for external storage, since USB and thunderbolt are faster, but it can be found in some CE devices, such as set-top boxes.

URL: www.sata-io.org

11.1.4 Thunderbolt

Thunderbolt is a brand name for a hardware interface developed by Apple and Intel that allows connection of external peripheral devices to a computer. Thunderbolt 1 and 2 used the same connector as Mini DisplayPort (MDP), while Thunderbolt 3 uses USB Type-C connectors.

The Thunderbolt interface combines PCI Express (PCIe) and DisplayPort (DP) into two serial signals and also provides DC power in a single cable. Up to six peripherals may be supported by one connector via hubs or daisy chains. The Thunderbolt controllers multiplex one or more individual data lanes from connected PCIe and DisplayPort devices for transmission via two duplex lanes and then demultiplexes them for use by PCIe and DisplayPort devices on the other end.

Intel introduced Light Peak at the 2009 Intel Developer Forum (IDF), using a prototype Mac Pro logic board running two 1080p video streams plus LAN and storage devices over a single 30-meter optical cable with modified USB ends. Intel switched to electrical connectors to reduce costs and to supply up to 10 W of power to connected devices. In February 2011, Apple introduced a new line of MacBook Pro notebook computers and announced that the technology's commercial name was Thunderbolt.

In June 2013 Intel announced that the next generation of Thunderbolt, Thunderbolt 2, is based upon the Falcon Ridge controller running at 20 Gbps (2.5 GB/s). Thunderbolt 2 incorporates DisplayPort 1.2 support, allowing video streaming to a single 4K video monitor. Thunderbolt 2 is backward compatible with Thunderbolt 1.

Thunderbolt 3 was developed by Intel and uses USB Type-C connectors. The Alpine Ridge Thunderbolt 3 controller provides a bandwidth of 40 Gbps (5 GB/s), halves power consumption from the prior generation, and can drive two external 4K displays at 60 Hz or a single external 4K display at 120 Hz. The controller supports PCIe 3.0, HDMI 2.0, and DisplayPort 1.2. Thunderbolt 3 cables can also source or sink up to 100 watts of power. This allows eliminating a separate power cable on some devices.

URL: https://thunderbolttechnology.net

11.1.5 NVMe

NVM Express (NVMe) is an open collection of standards and information to fully expose the benefits of nonvolatile memory in all types of computing environments. The NVM Express Work Group was incorporated as NVM Express, Inc. in 2014. The organization has over 100 member companies, including all the major solid-state drive manufacturers. Version 1.0 of the specification was released in October 2012. Version 1.2 was released in November 2014.

M.2, and the device form factor, formerly known at the Next Generation Form Factor (NGFF), is a specification for internally mounted computer expansion cards. It replaced the mSATA standard that used the PCI Express Mini Card layout and connectors. M.2 allows different module widths and lengths and, with advanced interfaces such as NVMe, makes this a good interface for solid-state drives in devices such as tablet computers. M.2 NVMe utilizes the capability of high-speed PCIe busses that are common in consumer mobile computers.

URL: http://www.nvmexpress.org

11.1.6 UFS

Universal flash storage or UFS is a flash memory storage interface specification for digital cameras, mobile phones, and other consumer electronic devices. UFS is positioned as a replacement for eMMCs and SD cards. The electrical interface for UFS using M-PHY is developed by the MIPI alliance, a high-speed serial interface targeting 2.9 Gbps per lane with a known path to 5.8 Gbps per lane. UFS is based upon the SCSI architectural model and supports SCSI Tagged Command Queuing. This is a JEDEC standard. Below is information about the various versions of UFS.

UFS	1.0	1.1	2.0	2.1
Introduced	24–02-2011[1]	25–06-2012[2]	18–09-2013[3]	04–04-2016[4]
Bandwidth per lane	300 MB/s		600 MB/s	
Max. number of lanes	1		2	
Max. total bandwidth	300 MB/s		1200 MB/s	
M-PHY version			3.0	
UniPro version			1.6	

URL: https://ufsa.org
[1]Cho, Hee Chang (August 2016). "Next Generation of Mobile Storage: UFS and UFS Card" (PDF). Jedec
[2]Chen, Horace (July 2017). "UFS 3.0 Controller Design Considerations" (PDF). Jedec
[3]"JEDEC Announces Publication of Universal Flash Storage (UFS) Standard | JEDEC". www.jedec.org. Retrieved 2017-05-08
[4]"JEDEC Updates Universal Flash Storage (UFS) Standard JEDEC". www.jedec.org. Retrieved 2017-05-08

11.1.7 Open NAND Flash (ONFI) Standard

The *Open NAND Flash Interface Working Group*, or *ONFI*, is a consortium of technology companies working to develop open standards for NAND flash memory chips and devices that communicate with them. The formation of ONFI was announced at the Intel Developer Forum in March 2006. The group's goals notably did *not* include the development of a new consumer flash memory card format. Rather, ONFI sought to standardize the low-level interface to raw NAND flash chips.

The ONFI consortium is led by several prominent manufacturers of NAND flash memory: Hynix, Intel, Micron Technology, Phison, SanDisk (now part of WD), Sony, and Spansion (now part of Cypress). ONFI produced a specification for a standard interface to NAND flash chips. Version 1.0 of this specification was released on December 28, 2006. It specifies:

- A standard physical interface (pinout) for NAND flash in TSOP-48, WSOP-48, LGA-52, and BGA-63 packages
- A standard mechanism for NAND chips to identify themselves and describe their capabilities (comparable to the serial presence detection feature of SDRAM chips)
- A standard command set for reading, writing, and erasing NAND flash
- Standard timing requirements for NAND flash
- Improved performance via a standard implementation of read cache and increased concurrency for NAND flash operations
- Improved data integrity by allowing optional ECC memory features

Version 2.0 that was released in February 2008 defined an interface for rates greater than 133 MB/s, whereas the legacy NAND interface was limited to 50 MB/s.

Version 2.1 that was released in January 2009 added features for higher rates of 166 MB/s and 200 MB/s, plus other enhancements to increase power, performance, and ECC capabilities. A verification product was announced in June 2009.

Version 2.2 released on October 2009 added: Individual LUN reset, enhanced program page register clear, and new ICC specs and measurement LUN reset and page register clear enable more efficient operation in larger systems with many NAND devices, while the standardized ICC testing and definitions will provide simplified vendor testing and improved data consistency.

The NAND Connector Specification was ratified in April 2008. It specifies a standardized connection for NAND modules (similar to DRAM DIMMs) for use in applications like caching and solid-state drives (SSDs) in PC platforms.

Version 2.3 was published in August 2010. It included a protocol called EZ-NAND that hid ECC details.

Version 3.0, published in March 2011, promoted a high-speed NAND Flash interface supporting transfer rates up to 400 MB/s. It required fewer chip-enabled pins enabling more efficient printed circuit board routing. A standard developed jointly with the JEDEC was published in October 2012.

Version 3.2, published on July 23, 2013, raised the data rate to 533 MB/s.

Version 4.0, published on April 17, 2014, introduced the NV-DDR3 interface increases the maximum switching speed from 533 MB/s to 800 MB/s, providing a performance boost of up to 50% for high-performance applications enabled by solid-state NAND storage components.

URL: www.onfi.org

11.1.8 Flash Card Standards

There are many types of removable storage cards used in consumer devices. These usually use flash memory although other types of memory have been used in these devices such as 1-inch hard disk drives in the CompactFlash form factor. Following is information on some of the most used flash card standards[1]:

Name	Acronym	Form factor
PC Card (www.pcmcia.org)	PCMCIA	85.6 × 54 × 3.3 mm
CompactFlash I (www.compactflash.org)	CF-I	43 × 36 × 3.3 mm
CompactFlash II (www.compactflash.org)	CF-II	43 × 36 × 5.5 mm
CFast (http://www.compactflash.org/cfast-cards)	CFast	43 × 36 × 3.6 mm
Multimedia Card (https://en.wikipedia.org/wiki/MultiMediaCard)	MMC	32 × 24 × 1.5 mm
Reduced Size Multimedia Card (https://en.wikipedia.org/wiki/MultiMediaCard)	RS-MMC	16 × 24 × 1.5 mm
MMCmicro Card (www.mmca.org)	MMCmicro	12 × 14 × 1.1 mm
Secure Digital Card (www.sdcard.org)	SD	32 × 24 × 2.1 mm
MiniSD Card (www.sdcard.org)	miniSD	21.5 × 20 × 1.4 mm
MicroSD card (www.sdcard.org)	microSD	11 × 15 × 1 mm
xD-Picture Card (www.xd-picture.com)	xD	20 × 25 × 1.7 mm

[1] http://en.wikipedia.org/wiki/Memory_card

Trusted Computing Group Standards

The *Trusted Computing Group (TCG)* is a nonprofit industry organization formed to develop, define, and promote open standards for hardware-enabled trusted computing and security technologies. The storage working group in the TCG focuses on standards for security services on dedicated storage systems. The SWG of the TCG develops standards and practices for defining the same security services across dedicated storage controller interfaces, including but not limited to ATA, Serial ATA, SCSI, Fibre Channel, USB storage, IEEE 1394, network-attached storage (TCP/IP), NVM Express, and iSCSI. Storage systems include disk drives, removable media drives, flash storage, and multiple storage device systems.

The TCG PC-based Trusted Platform Module (TPM) and the Trusted Software Stack (TSS) are designed into integrated circuits, systems, and applications. The TPM chips can be used for device authentication, rogue software detection, and secure storage. A key element in providing this security is encryption of data on the storage device with the encryption key kept in a nonuser accessible area on the storage device. There are three main security benefits of these specifications:

1. It creates trust relationships between storage devices and hosts using mutual identity and authentication of hosts and storage devices. By extending TPM into storage devices, the device can limit who can read or write to the storage device.
2. TCG-enabled storage enables secure control over storage device features. These products provide protected storage for specific users, systems, and applications. They also provide exclusive control of encrypted data on the TCG-enabled storage device.
3. They enable secure communications between storage devices and their hosts using session-oriented security commands defined in security extensions of storage device command specifications such as SCSI and ATA.

Since data at rest resides mostly on storage devices, these are a natural target for security. By encrypting the data and storing the key on a storage device, it becomes impossible to access this data through probe points and other weak points on encryption systems where the key is accessed off the disk drive. Also by incorporating encryption on the drive, no additional hardware or software is required with the attendant risk that that hardware or software could fail. TPM is enabled by a tiny processor that can sign and whose firmware cannot be modified.

Security services are called through specific APIs that protect the storage device behind a trust boundary. Only trusted entities with the proper authorization to the API can use TCG trusted storage. The protection of encrypted content on a storage device can be used to provide secure backup of content between different TCG-enabled storage devices. The TCG specification is also able to perform pervasive logging, meaning that it can support logging and clocking capabilities. This can help in determining when data should be moved from one device to another based upon use.

URL: https://trustedcomputinggroup.org/work-groups/storage/

11.2 Consumer Product Standards

There are several important standards covering the operation of consumer products. We present here current and some relatively recent but now defunct standards here (that may still be found in consumer devices).

There are also a great many options being developed for networking in the home that are used for sharing content between storage devices. We devote a section of this chapter on the viable home network standards.

11.2.1 UHAPI

NXP Semiconductors (formerly the semiconductor unit of Royal Philips Electronics) and the semiconductor unit of Samsung Electronics Co., Ltd. partnered to solve the task-level complexities that stall innovation and progress at all layers of the home CE stack. As a result, the *Universal Home API* (UHAPI) Forum was established to further this vision. The UHAPI Forum developed A/V APIs for home consumer electronic products such as DVD players and recorders, digital televisions, media servers, media adapters, and set-top boxes. The hardware-independent UHAPI enables independent software vendors, system integrators, and CE product manufacturers to create middleware and application software that is easily ported across multiple UHAPI-compliant solutions.

Last specification is UHAPI 1.2.

URL: http://elinux.org/UHAPI

11.2.2 DLNA

The *Digital Living Network Alliance* (DLNA) was formed in June 2003 by a group of consumer electronic companies, to develop and promote a set of interoperability guidelines for sharing digital media among multimedia devices under a certification standard. DLNA works with cable, satellite, and telecom service providers to provide link protection on each end of the data transfer. The extra layer of digital rights management (DRM) security allows broadcast operators to enable consumers to share their content on multimedia devices without the risk of piracy. As of June 2015, the organization claimed 200 member companies.

The group published its first set of guidelines in June 2004. The guidelines incorporate several existing public standards, including Universal Plug and Play (UPnP) for media management and device discovery and control and widely used digital media formats and wired and wireless networking standards.[6]

In March 2014, DLNA publicly released the VidiPath Guidelines, originally called “DLNA CVP-2 Guidelines.” VidiPath is a set of guidelines developed by

DLNA that enables consumers to view subscription TV content on a wide variety of devices including televisions, tablets, phones, Blu-ray players, set-top boxes (STBs), personal computers (PCs), and game consoles without any additional intermediate devices from the service provider.

As of September 2014, over 25,000 different device models had obtained "DLNA Certified" status, indicated by a logo on their packaging and confirming their interoperability with other devices. It was estimated that by 2017 over 6 billion DLNA-certified devices, from digital cameras to game consoles and TVs, would be installed in users' homes.

On January 5, 2017, DLNA announced on its web site that "the organization has fulfilled its mission and will dissolve as a non-profit trade association." Its certification program will be conducted by SpireSpark International of Portland, Oregon.

URL: https://www.dlna.org

11.2.3 OSGi Alliance

The *OSGi Alliance*, formerly known as the Open Service Gateway Initiative (OSGi), is an open standards organization founded in March 1999 that originally specified and continues to maintain the OSGi standard. The OSGi specification implements a complete and dynamic component model for the Java programming language. The original focus was on service gateways, but the specification is now used in applications including mobile phones, automobiles, and entertainment servers.

URL: www.osgi.org

11.2.4 Some Additional DRM Standards

MagicGate is a copy-protection technology introduced by Sony in 1999 as part of the Secure Digital Music Initiative (SDMI). It works by encrypting the content on the device and using MagicGate chips in both the storage device and the reader to enforce control over how files are copied.

MagicGate encryption was used in the memory cards of the PlayStation 2 and, as of 2004, has been introduced into all of Sony's Memory Stick products. Some devices, such as Sony's Network Walkman, will only accept Memory Sticks which use MagicGate technology. Note that MagicGate is not supported by many CE companies, and even many Sony products don't recognize cards with the MagicGate chips.

Content protection for recordable media and prerecorded media (CPRM/CPPM) is a mechanism for controlling the copying, moving, and deletion of digital media on a host device, such as a personal computer or other digital player. It was developed and controlled by the 4C Entity, LLC (which consists of IBM, Intel, Matsushita, and Toshiba).

The CPRM/CPPM specification is designed to balance the robustness and renewability requirements of content owners with ease of implementation for consumer electronic designers. The system defined by the specification relies on key management for interchangeable media, content encryption, and media-based renewability.

The CPRM/CPPM specification defines a cryptographic method for protecting entertainment content recorded on physical media. The types of physical media supported include recordable DVD media and flash memory. The most widespread use of CPRM is probably in Secure Digital cards.

11.3 Home Networking Standards

Following is a listing of various home wired and wireless networking standards with URLs where the reader can get more information.

11.3.1 Bluetooth

Bluetooth wireless technology is a short-range communication technology intended to replace the cables connecting portable and/or fixed devices while maintaining high levels of security. The key features of Bluetooth technology are robustness, low power, and low cost. The Bluetooth specification defines a uniform structure for a wide range of devices to connect and communicate with each other.

Latest core specification is v5.0 adopted in December 2016.

URL: www.bluetooth.com

11.3.2 CableHome

CableHome developed the interface specifications necessary to extend high-quality cable-based services to network devices within the home. It was sponsored by CableLabs. The objective of the CableHome™ architecture is to provide a variety of new services to residential customers, simplify the management the home network, and protect copyrighted information from being diverted to other uses. Within the residence, the CableHome architecture is designed to be independent of the underlying network technologies such as Ethernet, wireless, HomePNA, and HomePlug.

CableHome 1.1 specification was released in July 2006.

11.3.3 DOCSIS

DOCSIS, the Data Over Cable Service Interface Specification, is originally developed at CableLabs. It defines interface requirements for cable modems involved in high-speed data distribution over cable television system networks. DOCSIS certification provides cable modem equipment suppliers with a fast, market-oriented method for attaining cable industry acknowledgment of DOCSIS compliance and is responsible for high-speed modems being certified for retail sale.

The latest interface specification is DOCIS 3.1 with downstream and upstream capacity of 10 Gbps at full duplex.

URL: http://www.cablelabs.com/full-duplex-docsis/

11.3.4 IEEE 1394

IEEE 1394 multimedia connection enabled simple, low-cost, high-bandwidth isochronous (real-time) data interfacing between computers, peripherals, and consumer electronic products such as camcorders, VCRs, printers, PCs, TVs, and digital cameras. With IEEE 1394-compatible products and systems, users transferred video or still images from a camera or camcorder to a printer, PC, or television, with no image degradation. IEEE 1394 is not being built into new consumer products.

URL: www.1394ta.org/Technology/index.htm

11.3.5 WirelessHD, WHDI, and WiGig

WirelessHD or *WiHD* was a special interest group formed to develop a specification for a wireless high-definition digital interface (originally developed by Silicon Image) that enables high-definition audio/video (A/V) streaming and high-speed content transmission for consumer electronic (CE) devices. This is an in-room solution running at 60 GHz. It uses a steerable antenna array to form a beam that can transmit content up to 500 MBps within a 10 m distance using the 7 GHz channel. This more or less line-of-sight technology is intended to allow devices such as DVD players or DVRs to move HD content to a monitor without the use of HDMI wiring. In operation WirelessHD needs to have a line of sight connection to work well.

WHDI operates at 5 GHz with a 30 m maximum range. WHDI doesn't suffer from the same interference issues that WirelessHD has and thus doesn't require line of sight to work. However there is less use of WHDI than WirelessHD in actual products.

WiGig operates at 60 GHz and is a line-of-sight, in-room technology. WiGig is also known as 802.11ad, discussed below.

There are also a number of technologies that use local Wi-Fi for wireless streaming such as AirPlay, Chromecast, and Miracast.

URLs:

WirelessHD: https://en.wikipedia.org/wiki/WirelessHD
WHDI: http://www.whdi.org
WiGig: http://www.wi-fi.org/discover-wi-fi/wi-fi-certified-wigig

11.3.6 IEEE 802

IEEE 802 standard comprises a family of networking standards that cover the media access control (MAC) and physical layer (PHY) specifications of technologies from Ethernet to wireless. IEEE 802 is subdivided into 25 parts that cover the physical and data-link aspects of networking. The better-known specifications include 802.3 Ethernet, 802.11 Wi-Fi, 802.15 Bluetooth/ZigBee, and 802.16.

The following table lists highlights of the most popular sections of IEEE 802 as of the time of this writing.[2] Copies of these specifications are available from the IEEE. The most popular specifications are in bold.

802	Overview	Basics of physical and logical networking concepts
802.1	Bridging	LAN/MAN bridging and management. Covers management and the lower sub-layers of OSI Layer 2, including MAC-based bridging (media access control), virtual LANs, and port-based access control
802.3	**Ethernet**	"Grandaddy" of the 802 specifications. Provides asynchronous networking using "carrier sense, multiple access with collision detect" (CSMA/CD) over coax, twisted pair copper, and fiber media. Current speeds range from 10 Mbps to 10 Gbps
802.11	**Wi-Fi**	Wireless LAN media access control and physical layer specification. 802.11a, b, g, etc. are amendments to the original 802.11 standard. Products that implement 802.11 standards must pass tests and are referred to as "Wi-Fi certified"
802.11a		Specifies a PHY that operates in the 5 GHz U-NII band in the USA—initially 5.15–5.35 AND 5.725–5.85—since expanded to additional frequencies Uses orthogonal frequency-division multiplexing Enhanced data speed to 54 Mbps Ratified *after* 802.11b
802.11b		Enhancement to 802.11 that added higher data rate modes to the DSSS (Direct Sequence Spread Spectrum) already defined in the original 802.11 standard Boosted data speed to 11 Mbps 22 MHz Bandwidth yields 3 non-overlapping channels in the frequency range of 2.400 GHz to 2.4835 GHz Beacons at 1 Mbps, falls back to 5.5, 2, or 1 Mbps from 11 Mbps max.
802.11d		Enhancement to 802.11a and 802.11b that allows for global roaming Particulars can be set at media access control (MAC) layer

(continued)

[2] https://en.wikipedia.org/wiki/IEEE_802

802	Overview	Basics of physical and logical networking concepts
802.11e		Enhancement to 802.11 that includes quality of service (QoS) features Facilitates prioritization of data, voice, and video transmissions
802.11g		Extends the maximum data rate of WLAN devices that operate in the 2.4 GHz band, in a fashion that permits interoperation with 802.11b devices Uses OFDM modulation (orthogonal FDM) Operates at up to 54 megabits per second (Mbps), with fallback speeds that include the "b" speeds
802.11h		Enhancement to 802.11a that resolves interference issues Dynamic frequency selection (DFS) Transmit power control (TPC)
802.11i		Enhancement to 802.11 that offers additional security for WLAN applications Defines more robust encryption, authentication, and key exchange, as well as options for key caching and pre-authentication
802.11j		Japanese regulatory extensions to 802.11a specification Frequency range 4.9 GHz to 5.0 GHz
802.11k		Radio resource measurements for networks using 802.11 family specifications
802.11m		Maintenance of 802.11 family specifications Corrections and amendments to existing documentation
802.11n	MIMO_OFDM	Higher-speed standards—up to 150 Mbps 4 MIMO streams Up to 70 m indoor range
802.11ac	MIMO_OFDM	Up to 867 Mbps 8 MIMO streams Up to 35 m indoor range
802.11ad	OFDM	WiGig used for media streaming over a 60 GHz line-of-sight path within a room
802.11x		Misused "generic" term for 802.11 family specifications
802.15	**Wireless PAN**	Communication specification that was approved in early 2002 by the IEEE for wireless personal area networks (WPANs)
802.15.1	**Bluetooth**	Short-range (10 m) wireless technology for cordless mouse, keyboard, and hands-free headset at 2.4 GHz
802.15.3a	UWB	Short-range, high-bandwidth "ultra-wideband" link
802.15.4	**ZigBee, WirelessHART, MiWi, etc.**	Short-range wireless sensor networks
802.15.5	Mesh network	Extension of network coverage without increasing the transmit power or the receiver sensitivity Enhanced reliability via route redundancy Easier network configuration—better device battery life
802.15.6	Body area network	For communication around and using a human body

(continued)

802	Overview	Basics of physical and logical networking concepts
802.16	Wireless metropolitan area networks	This family of standards covers Fixed and Mobile Broadband Wireless Access methods used to create wireless metropolitan area networks (WMANs). Connects base stations to the Internet using OFDM in unlicensed (900 MHz, 2.4, 5.8 GHz) or licensed (700 MHz, 2.5–3.6 GHz) frequency bands. Products that implement 802.16 standards can undergo *WiMAX* certification testing
802.18	Radio Regulatory TAG	IEEE 802.18 standards committee
802.19	Coexistence	IEEE 802.19 Coexistence Technical Advisory Group
802.21	Media Independent Handoff	IEEE 802.21 mission and project scope
802.22	Wireless regional area network	IEEE 802.22 mission and project scope
802.24	Smart Grid TAB	For wireless management of power sources and sinks
802.25	Omni-Range Area Network	For broader area wireless communication

CableLabs Video Specification

The *CableLabs Video Specification*, formerly known as the OpenCable initiative, covers hardware, software, and middleware used to deliver video content on a cable system. This includes linear, VOD, interactive advertising, IP delivery, home networking, and related applications. Tru2way is the consumer facing brand for these services.

OpenCable uses SCTE standards for the video, transport, and various interface requirements but also adds a requirement for a Java-based software interpreter to support the OpenCable Application Platform (OCAP). It also requires a decryption system for protected content employing CableCARDs or the proposed software-based downloadable conditional access system (DCAS).

URL: http://www.cablelabs.com/specifications-library/opencable/

11.3.7 PacketCable

PacketCable is a CableLabs-led initiative built on the DOCSIS cable modem infrastructure. This is an interoperable interface specifications for delivering advanced, real-time multimedia services over two-way cable. PacketCable networks use Internet Protocol (IP) technology to enable a wide range of multimedia services, such as IP telephony, multimedia conferencing, interactive gaming, and general multimedia applications.

PacketCable 2.0 is the most recent release.

URL: http://www.cablelabs.com/specifications-library/packetcable/

11.3.8 Voice Over IP

Voice over Internet Protocol is also called VoIP. Other names for this technology are IP telephony, Internet telephony, broadband telephony, and voice over broadband. This technology involves the routing of voice conversations over the Internet or though other IP-based networks.

URL: www.tiaonline.org/standards/technology/voip

11.3.9 Universal Plug and Play

The *Universal Plug and Play (UPnP) architecture* offers pervasive peer-to-peer network connectivity of PCs of all form factors, intelligent appliances, and wireless devices. The UPnP architecture is a distributed, open networking architecture that leverages TCP/IP and the web to enable seamless proximity networking in addition to control and data transfer among networked devices in the home, office, and everywhere in between. UPnP provides the following capabilities:

- *Media and device independence.* UPnP technology can run on any network technology including Wi-Fi, coax, phone line, power line, Ethernet, and 1394.
- *Platform independence.* Vendors can use any operating system and any programming language to build UPnP products.
- *Internet-based technologies.* UPnP technology is built upon IP, TCP, UDP, HTTP, and XML, among others.
- *UI Control.* UPnP architecture enables vendor control over device user interface and interaction using the browser.
- *Programmatic control.* UPnP architecture enables conventional application programmatic control.
- *Common base protocols.* Vendors agree on base protocol sets on a per-device basis.
- *Extendable.* Each UPnP product can have value-added services layered on top of the basic device architecture by the individual manufacturers.

URL: https://en.wikipedia.org/wiki/Universal_Plug_and_Play

11.4 Needed Standards for Future Consumer Electronic Development

Following are some ideas for standard that we will need to carry about some of the initiatives described in this book. These range from creating applications integrated into storage devices to creating the integrated home storage utility.

11.4.1 Proposal for Open Standards for Storage Integration Into Consumer Electronics

The author believes that the best approach to facilitate the development of integrated applications in digital storage devices is to create a consortium to develop open integration standards that allow both the consumer electronic companies and the storage device manufacturers to maintain the means of creating proprietary products while improving the cost and performance of these products and increasing the ease and speed with which consumer end products can be created. A useful standard should also effectively deal with the issues of control of the end product and its functionality to the mutual satisfaction of the consumer electronic company and the digital storage device company.

Development and implementation of such a standard could be beneficial to the host company by allowing rapid product design with better performance and lower overall cost. It would be beneficial to the storage device companies by allowing them to increase the value add of their products and thus improve their profit margins.

Following is a proposal and some expectations for the creation of such an open standard for consumer application integration into storage devices:

(a) The basic idea is to develop an open standard for the design, testing, and manufacturing of a target storage device, such as a hard disk drive or an SSD with integrated electronic application modules contained on in the storage device electronics. These devices can be controlled by firmware and appropriate interfaces with the outside environment to create customizable proprietary consumer electronic applications.

(b) The standard must allow for the updating and importation of firmware for controlling the resulting applications or for reconfiguring between applications when appropriate.

(c) The standard must create flexible and yet defined rules for the incorporation of standard consumer application interfaces on the circuit board of the storage device to connect with the application/protective housing.

(d) This approach should save considerably on the bill of materials costs of the resulting consumer electronic devices.

(e) This approach should also save considerably on the manufacturing costs of the resulting consumer electronic devices through the use of a single manufacturing line and an integrated drive and application test.

(f) In addition, tight integration of the storage device with the host application should allow considerable improvement in host device performance as well as improvement in such factors such as power usage.

(g) This standard could also include other ancillary capabilities that could be incorporated into the integrated consumer applications such as application SMART that can log application performance in a storage device log file and flag potential failures of the application modules.

11.4.2 Standards for Personal Content Metadata and Organization of Personal Content

As our personal content grows, it will quickly become unmanageable unless we develop ways to automatically organize and manage that data. For this reason, we need to develop metadata format standards and processes for the automatic generation of metadata into a universal format. These are needed because for personal content most users don't have the time to organize the digital files that result from living their lives. These standards should be incorporated as much as possible with the applications that use the content in order to create a process that works seamlessly with the applications.

Likewise, we need search and indexing capabilities built either into applications or in a location, like an integrated home network or cloud services, where this organization and management takes place. It could be that applications in static or mobile devices do the initial metadata capture and organization with deep indexing and associations created on a centralized and more powerful computer like that in a centralized home server appliance or in a data center through the cloud. Once organized, the metadata and proxies of the content can be loaded back on the mobile devices so that they can act as a personal memory assistant. These functions would also benefit from established metadata standards.

In addition to standards, if personal content, particularly life logs, increase at the rate projected in this book, we will need legal safeguards to ensure the privacy of this personal content. Perhaps privacy can be dealt with using encryption technologies. In addition, should this personal content be open to court orders or could a user with such content plead the Fifth Amendment?

11.4.3 Standards for Virtualization of Consumer Storage and the Creation of a Home Storage Utility

At the enterprise storage level, the Storage Networking Industry Association (SNIA) has created standardization of storage management. SNIA's Storage Management Initiative Specification (SMI-S) has been widely adopted as a storage resource management standard and has laid the foundation for management software to be able to be written to one set of interfaces defined by an SMI-S profile for each type of device.

As the complexity of devices with storage as well as the sheer volume of digital content in the home increases, automated management processes will be needed. It is likely that some of the standards and practices developed initially to support enterprise network storage will eventually migrate to support networked devices with storage in the home.

Virtualization of enterprise computer and storage assets has improved storage device utilization. Disk storage utilization in the enterprise used to be 50% or less.

Using virtualization of available storage in storage arrays into a common managed pool combined with technologies such as deduplication, it is possible to increase overall disk utilization to greater than 70%. This lowers the cost of maintaining the data center utility (for instance, fewer storage arrays means less heat generated which air conditioning has to remove) as well as lowers the total storage device costs.

In the home, some sort of virtualization or abstraction of the storage in all the devices in and around the home with storage and connected to the network or in the cloud could allow more effective management of both the storage resources and personal and commercial content. This unified management would allow efficient backup of content, moving content between storage devices and making the most effective use of the network bandwidth to provide content where it is needed.

The storage networking standards developed by organizations such as SNIA combined with special capabilities for the unique features of the home storage environment will help bring a unified home storage pool or utility to fruition. Development of the metadata standards and automated generation of metadata will also help in content management. These standards will help the system determine where particular content should go and who should have access to it. Also, these capabilities could enable entertainment using the home network that includes personal as well as commercial content.

Appendix A. Home Networking Technology Trade Groups

The *HomePlug Alliance* was founded to support data transport over power lines. Initial Alliance members include Panasonic, Sharp Corp., Conexant Systems Inc., and Intellon Corporation. Sustained data rate reported for the HomePlug AV specification allows for 200 Mbps data rates or about 25 MB/s. There are 35 member companies and over 200 certified products.

URL: www.homeplug.org

The *HomePNA* standard was developed by the *Home Phoneline Networking Alliance*. It was founded in June 1998 and is led by seven companies (Agere Systems, AT&T, Broadcom, Conexant, Hewlett-Packard Co., Motorola, and 2Wire). HomePNA 2.0 was developed to run over existing phone line and coaxial wiring in a symmetrical mode at a peak data rate up to 32Mbps, with an average throughput of 10 Mb/s (~1.25 MB/s). The HPNA 2.0 network uses a shared physical media (wiring) has no need for a switch or hub. HomePNA merged with HomeGrid in 2013.

URL: www.homepna.org

The Multimedia over Coax Alliance or MoCA was established in 2004 and is an international standards consortium publishing specifications for networking over coaxial cable. There are three basic versions of the specification currently available, MoCA 1.1, MoCA 2.0, and MoCA 2.5. Following is an outline of characteristics of these specifications.

URL: www.mocalliance.org

	MoCA 1.0	MoCA 1.1	MoCA 2.0	MoCA 2.0bonded	MoCA 2.1	MoCA 2.1bonded	MoCA 2.5
Mbit/s actual throughput	100	175	500	1000	500	1000	2500
Number of channels bonded				2		2	3~5
Power save (standby and sleep)			X	X	X	X	X
MoCA protected setup (MPS)					X	X	X
Management proxy					X	X	X
Enhanced privacy					X	X	X
Network wide beacon power					X	X	X
Bridge detection					X	X	X

Appendix B. Companies Making Various Storage Products Used in Consumer Applications

Flash Memory Manufacturers[3]

There are many manufacturers of flash memory used in consumer applications. These companies are located in many locations although the largest volume of flash memory is produced in Asia. Following is a list of many of the major suppliers of flash memory:

- *Adesto* offers parallel flash memories in a variety of configurations and architectures through its acquisition of Atmel's nonvolatile memory business. Adesto's flash memory is used in PC peripherals, mobile phones, networking equipment, and set-top boxes. *URL:* http://www.adestotech.com/
- *Cypress Semiconductor* merged with Spansion to make it one of the leading companies making SRAM, NOR, and SLC NAND flash memory. URL: http://www.cypress.com
- *GigaDevice* is a company in Beijing providing SRAM and SPI NOR flash memory. URL: http://www.gigadevice.com/index.html

[3] Jim Handy of Objective Analysis helped with the update on the flash memory companies.

- *Intel* makes NAND flash devices with multilevel cells in planar and increasingly 3D cell configurations. Intel has joint flash memory fabs with Micron (IMFT) and its own fabs in China. Intel and Micron have jointly developed the 3D XPoint memory technology. *URL:* www.intel.com/
- *ISSI (Integrated Silicon Solution, Inc.)* manufactures NOR flash chips for embedded applications. *URL:* http://www.issi.com/
- *Macronix International* is a leading supplier of read-only memories (ROM) as well as advanced NOR memories. It also sells legacy NAND solutions. URL: http://www.macronix.com/en-us/
- *Micron* makes DRAM and NOR flash memory as well as NAND flash devices with some unique capabilities. It makes planar and 3D multiple bit per cell flash memory using floating gate technology. Micron Flash memory also offers high-performance features, including a *read cache* function that streams more than 30 MB per second. The company has joint NAND production facilities with Intel and codeveloped the 3D XPoint memory technology. *URL:* www.micron.com/
- *Samsung* is the largest manufacturer of NAND flash memory in the world. It was the first to introduce 3D NAND memory, and most of its NAND flash production uses this technology. Samsung flash is used in many mobile consumer applications. *URL*: http://www.samsung.com/semiconductor/products/flash-storage/
- *SK hynix* in Korea is one of the world's largest NAND flash manufacturers. The company is owned by SK Telecom, Korea's leading cell phone service provider. *URL:* http://www.skhynix.com/eng/product/nandRaw.jsp
- *Toshiba* invented flash memory and remains a leader in this market. They produce a great variety of flash memory including planar and 3D NAND flash. Toshiba makes SSDs as well as flash cards. They jointly own flash production facilities with SanDisk (now Western Digital). *URL:* https://toshiba.semicon-storage.com/us/product/memory.html
- *Western Digital (formerly SanDisk)* was an early pioneer in flash memory and the inventor of flash memory cards. SanDisk has the biggest retail presence in the card memory market, especially in the USA. SanDisk manufactures its flash memory in factories jointly owned with Toshiba (although at the time of this writing Toshiba was selling its memory unit). *URL:* www.sandisk.com
- *Winbond* is a Taiwan-based IC company that makes low-density NOR flash memories for the consumer electronic and communication markets. URL: http://www.winbond.com/hq?__locale=en

Hard Disk Manufacturers

The hard disk drive industry has undergone significant consolidation over the years. Where there once were close to 50 companies making hard disk drives, there are now only 3. The listing of products and markets in the descriptions to follow are current as of the time of writing and will probably change with time.

- *Toshiba* is the smallest HDD company. It pioneered small form factor disk drives from 0.85-inch to 2.5-inch. They developed the 0.85-inch drive for the mobile phone market. The 0.85-inch to 1.8-inch products are no more and Toshiba now makes only 2.5-inch and 3.5-inch HDDs (the 3.5-inch products it acquired through a Fujitsu acquisition). *URL:* http://storage.toshiba.com/consumer-hdd
- *Seagate* is the second largest disk drive company. They make the full range of hard disk drive form factors. They are a significant supplier of HDDs for the CE industry. *URL:* www.seagate.com
- *Western Digital* is the largest hard disk drive company, making 3.5-inch and 2.5-inch hard disk drives used in computer and consumer applications. They have been a major supplier of HDDs for set-top boxes and game machines. *URL:* www.wdc.com

Optical Disk Manufacturers

Optical disk drives and media are manufactured by many companies. The actual manufacturing is generally in Asia, particularly in Taiwan or China. Some of the companies that make optical disk drives include Hitachi LG, Panasonic, Philips Lite-On, Pioneer, Sharp, and Sony.

External Storage Manufacturers

There are a great many companies making various external storage products with USB, Thunderbolt, or eSATA interfaces. Most of these use hard disk drives either singly or in small arrays, but some companies also supply external boxes with SSDs. Several of the hard disk drive manufacturers do this such as Hitachi, Seagate, Toshiba, and Western Digital.

Many other companies buy hard drives and add them to external enclosure boxes or make and sell these boxes to consumers to populate. Among the external storage suppliers without their own source of disk drives are Buffalo, HP, Iomega, LaCie, Drobo, and many others.

Bibliography

A Note on Sources

What follows are various sources that are used in the process of creating this book. Some sources were used throughout the book and hence they are listed under general sources. Others were useful for particular chapters, and they are listed by chapter.

General

S. Kipp, *Broadband Entertainment* (All Digital Publishing, Westminster, 2004)
A. Dhir, *The Digital Consumer Technology Handbook* (Elsevier Press, New York, 2004)
B. Haskell, *Portable Electronics Product Design and Development* (McGraw-Hill, New York, 2004)
A.S. Hoagland, J.E. Monson, *Digital Magnetic Recording* (Wiley Interscience, New York, 1991)

Chapter 2: Fundamentals of Hard Disk Drives

S.X. Wang, A.M. Taratorin, *Magnetic Information Storage Technology* (Academic Press, San Diego, 1999)
E.D. Daniel, C.D. Mee, M.H. Clark, *Magnetic Recording* (IEEE Press, Piscataway, 1999)
F. Schmidt, *The SCSI Bus and IDE Interface* (Addison-Wesley, Reading, 1997)

Chapter 3: Fundamentals of Optical Storage

J. Taylor, *Everything You Ever Wanted to Know About DVD* (McGraw-Hill, New York, 2004)
T.W. McDaniel, R. Victora, *Handbook of Magneto-Optical Data Recording: Materials, Subsystems, Techniques* (Noyce Publications, Palo Alto, 1997)

T.M. Coughlin, *Digital Storage in Consumer Electronics*,
https://doi.org/10.1007/978-3-319-69907-3

Chapter 4: Fundamentals of Flash Memory and Other Solid State Memory Technologies

B. Dipert, M. Levy, *Designing with Flash Memory* (Annabooks, San Diego, 1994)
N. Gershenfeld, *The Physics of Information Theory* (Cambridge University Press, New York, 2000)
R. Micheloni, *3D Flash Memories* (Springer, New York, 2016)

Chapter 5: Storage in Home Consumer Electronics Devices

C. S. Swartz (ed.), *Understanding Digital Cinema* (Elsevier, New York, 2005)
K. Jack, *Video Demystified* (Elsevier, New York, 2005)

Chapter 6: Storage in Automotive and Mobile Consumer Electronics Devices

L. Simone, *If I Only Changed the Software Why is the Phone on Fire?* (Elsevier, New York, 2007)
J. Handy, *The Cache Memory* (Academic Press, San Diego, 1993)
Y. Chen, L. Li, *Advances in Intelligent Vehicles* (Academic Press, San Diego, 2014)

Chapter 7: Developments in Mobile Consumer Electronic Enabling Technologies

R.R. Hainich, O. Bimber, *Displays: Fundamentals & Applications, Second Edition* (CRC Books, Boca Raton, 2017)
T. Reddy, *Linden's Handbook of Batteries, 4th Edition* (McGraw Hill, New York, 2011)
M. Lei Zeng, J. Qin, *Metadata, 2nd Edition* (ALA Neal-Schuman, Chicago, 2016)

Chapter 8: Integration of Storage in Consumer Devices

V.C. Hamacher, Z.G. Vranesic, S.G. Zaky, *Computer Organization* (McGraw Hill, New York, 1984)
N. Gershenfeld, *When Things Start to Think* (Henry Holt and Company, New York, 1999)
A. Bindal, *Electronics for Embedded Systems* (Springer, New York, 2017)

Chapter 9: Home Network Storage, The Cloud and the Internet of Things

A. Kovalick, *Video Systems in an IT Environment* (Focal Press, New York, 2006)
T. Clark, *Storage Virtualization* (Addison Wesley, Reading, 2005)
M. Farley, *Building Storage Networks* (McGraw-Hill/Osborne, New York, 2000)
R. Perlman, *Interconnections*, 2nd edn. (Addison-Wesley, Reading, MA, 1999)
D. Hanes, *IoT Fundamentals: Networking Technologies, Protocols and Use Cases for the Internet of Things* (Cisco Press, Indianapolis, 2017)

Chapter 10: The Future of Home Digital Storage

J.R. Pierce, *An Introduction to Information Theory* (Dover Publications, New York, 1980)
D. Norman, *The Design of Everyday Things* (Doubleday Currency, New York, 1990)
C. Anderson, *The Long Tail* (Hyperion, New York, 2006)
J. Duato, S. Yalamanchili, L. Ni, *Interconnection Networks, An Engineering Approach* (IEEE Computer Society, Piscataway, 1997)
F. Bunn, N. Simpson, R. Peglar, G. Nagle, *Storage Virtualization, SNIA Technical Tutorial* (Storage Networking Industry Association, Colorado Springs, 2004)
R. Cummings, *Storage Network Management, SNIA Technical Tutorial* (Storage Networking Industry Association, Colorado Springs, 2004)

Printed in the United States
By Bookmasters

Printed in the United States
By Bookmasters